Adaptive Middleware for the Internet of Things –
The GAMBAS Approach

RIVER PUBLISHERS SERIES IN COMMUNICATIONS

Indexing: All books published in this series are submitted to the Web of Science Book Citation Index (BkCI), to SCOPUS, to CrossRef and to Google Scholar for evaluation and indexing.

The "River Publishers Series in Communications" is a series of comprehensive academic and professional books which focus on communication and network systems. Topics range from the theory and use of systems involving all terminals, computers, and information processors to wired and wireless networks and network layouts, protocols, architectures, and implementations. Also covered are developments stemming from new market demands in systems, products, and technologies such as personal communications services, multimedia systems, enterprise networks, and optical communications.

The series includes research monographs, edited volumes, handbooks and textbooks, providing professionals, researchers, educators, and advanced students in the field with an invaluable insight into the latest research and developments.

For a list of other books in this series, visit www.riverpublishers.com

Adaptive Middleware for the Internet of Things –
The GAMBAS Approach

Marcus Handte

Universität Duisburg-Essen
Germany

Pedro José Marrón

Universität Duisburg-Essen
Germany

Gregor Schiele

Universität Duisburg-Essen
Germany

Manuel Serrano Matoses

Etra Investigación Y Desarrollo
Spain

River Publishers

Published, sold and distributed by:
River Publishers
Alsbjergvej 10
9260 Gistrup
Denmark

River Publishers
Lange Geer 44
2611 PW Delft
The Netherlands

Tel.: +45369953197
www.riverpublishers.com

ISBN: 978-87-9351-978-7 (Hardback)
      978-87-9351-977-0 (Ebook)

Contents

Preface

In early 2011, a small group of researchers and developers from three academic institutions, the Universität Dusiburg-Essen, the National University of Ireland in Galway and the Open University, and three commercial entities, ETRA I+D, EMT Madrid and Aristech, jointly decided to collaborate to address the challenges imposed by the impedance mismatch between information availability and access methods. After an initial set of discussions, they agreed on forming a research consortium that would extend the basic concepts of today's information systems in order to support their automatic adaptation to the context of their users. Instead of requiring users to provide large amounts of inputs to find and access a particular piece of information, the members of the consortium envisioned information services that would automatically offer the right information at the right time.

Enabling such a shift in system design required a new breed of information services. Instead of being driven by sequences of explicit user inputs, the goal was to have the services react to the behavior of their users. Towards this end, the services would need to access an up-to-date view of the users' context in order to adapt to their behavior automatically. Without adequate system support, this would lead to complex application logic that would have to capture, process and share large amounts of potentially private data. The resulting complexity for application developers would often outweigh the potential benefits of having behavior-driven services, especially, when considering small and medium-sized enterprises that could not afford the development of a powerful software infrastructure to support services.

To address this problem, the consortium applied for a research grant in the 7th Framework Programme of the European Union and acquired funding to design and implement GAMBAS, the **G**eneric **A**daptive **M**iddleware for **B**ehavior-Driven **A**utonomous **S**ervices. The hardware basis for this middleware was planned to be widely available personal mobile devices, such as smartphones and laptops, existing services on the Internet and upcoming Internet-connected Objects that would form the **I**nternet **of T**hings (IoT). Due to the composition of the consortium, which consisted of two companies in the public transport domain, the focus of the application areas described in the grant proposal were mobility and environmental monitoring applications in a smart city domain.

After the positive evaluation of the research grant, the consortium started its work on GAMBAS in February 2012. Three years later, the consortium successfully completed the project and provided not only a fully functional middleware system that was available under a public source license, but also developed a significant number of services and applications that demonstrated the maturity of the concepts and implementation. During the middleware and application development, the consortium published more than 25 articles and papers in conferences and journals and pushed the state of the art in adaptive data acquisition, interoperable data processing and privacy preservation.

Since 2011, the computing landscape has changed. IoT has made its way from academic conferences into mainstream and is present in the minds of regular people. Even small and medium-sized companies are manufacturing Internet-connected devices and processing personal information on a large scale. However, at the same time, we continuously hear new reports on data breaches that cause the release of large amounts of personal data from fitness trackers, IP-based home surveillance cameras and other inexpensive hardware capable of some sort of data acquisition. This shows that the concepts related to the dynamic and distributed processing of sensor information in a privacy-preserving manner realized by the GAMBAS middleware are not obsolete. Instead, they are more relevant today than they were seven years ago.

Looking at the latest consumer trends, we see that digital assistants are on the rise and many people are now installing Internet-connected microphones in their home environments to access them at any point in time. At a conceptual level, the idea of digital assistants can be seen as a specific implementation of the idea of autonomous behavior-driven services. By learning about the context of the user, e.g. through the user's phone book, calendar and the user's interaction with the assistant, the service continuously improves its accuracy and usefulness. Thereby, the service provides a easy-to-use user interface-based natural language understanding. However, when focusing on the details of the underlying implementations, we find rather closed systems operated exclusively by the largest players in the computing industry. In addition, the systems require users to put their full trust into their manufacturer.

In contrast, the idea behind behavior-driven services in GAMBAS is to avoid such single-points-of-trust by facilitating secure sharing and distributed processing of data based on an interoperable data representation. As a consequence, we are convinced that the concepts proposed in GAMBAS have not lost their appeal. Instead, we think that they are an important alternative to the centralized architectures that are in use today. By summarizing the GAMBAS approach to IoT middleware in this book, we hope to inspire future designers and developers to consider the concepts implemented in the GAMBAS middleware as design choices whose applicability has been demonstrated in several applications.

List of Figures

List of Abbreviations

AES	Advanced Encryption Standard
API	Application Programming Interface
ARM	Advanced RISC Machines
BESOS	Building Energy Decision Support Systems for Smart Cities
CCS	Constrained Computer System
CPU	Central Processing Unit
CQELS	Continuous SPARQL
CQP	Continuous Query Processor
DB	Database
DBMS	Database Management Sytem
DDR	Data Discovery Registry
DH	Diffie Hellman
DQF	Distributed Query Processing Framework
ECDH	Elliptic-Curve Diffie-Hellman
EDBC	Event-driven Backward Chaining
FFT	Fast Fourier Transform
FOAF	Friend of a Friend
GAMBAS	Generic Adaptive Middleware for Behavior-driven Autonomous Services
GPRS	General Packet Radio Service
GPS	Global Positioning System
GSM	Global System for Mobile Communications
GTFS	General Transit Feed Specification
ICT	Information and Communication Technology
IDE	Integrated Development Environment
IETF	Internet Engineering Task Force
ISO	International Organization for Standardization
IUI	Intent-aware User Interface
JSON	Javascript Object Notation
LAN	Local Area Network

LBS	Location-based Service
LOD	Linked Open Data
MAC	Message Authentication Code
MB	Megabyte
NFC	Near Field Communication
OBD	Onboard Diagnostics
OS	Operating System
OSI	Open Systems Interconnection
OWL	Web Ontology Language
PECES	Pervasive Computing in Embedded Systems
PIKE	Piggy-backed Key Exchange
PLANET	Platform for the Deployment and Operation of Heterogeneous Networked Cooperating Objects
PPO	Privacy Preference Ontology
PRF	Privacy Preservation Framework
QP	Query Processor
RAM	Random Access Memory
RDF	Resource Description Framework
RF	Radio Frequency
RSA	Rivest-Shamir-Adleman
SDK	Software Development Kit
SDS	Semantic Data Storage
SIMON	Assisted Mobility for Older and Impaired Users
SPARQL	SPARQL Protocol and RDF Query Language
SPI	Service Programming Interface
SPITFIRE	Semantic-Service Provisioning for the Internet of Things using Future Internet Research by Experimentation
SPT	Spitfire Ontology
SSL	Secure Sockets Layer
UI	User Interface
UMTS	Universal Mobile Telecomunication System
URI	Unified Resource Identifier
XML	Extensible Markup Language
YANTRIP	Yet Another N-Triple Parser

1

Introduction

This chapter first introduces the motivation behind the developments described in this book. Then, it discusses the main objectives of GAMBAS and describes the two motivating scenarios in the domain of mobility and environmental monitoring. Based on these scenarios, the chapter derives the overall vision and identifies the innovative characteristics. Finally, the chapter closes with a discussion of the state of the art that is used to highlight the primary innovations realized by the development of the GAMBAS middleware.

1.1 Motivation

With the advent of powerful personal mobile devices such as smart phones, digital assistants and tablet computers, an ever-increasing number of people has constant access to the wealth of information stored on the millions of servers connected via the Internet. Over the last years, the availability of such devices has caused a paradigm shift in the way people deal with information. Instead of collecting and printing potentially relevant documents in advance, using a personal computer that is only available at particular locations, they now access information on-demand and on-the-go.

Yet, despite this significant change in behavior, the technical means to access information have only changed marginally. As depicted in Figure 1.1, in most cases, information is accessed via the web, which requires users to memorize long URLs, click through sequences of web pages or browse irrelevant search results. Alternatively, if they are frequently accessing the same service, they may install an app or application that provides more convenient access. However, such an installation requires advance planning and does not provide suitable support for services that are primarily useful in a particular environment. Moreover, even if they are using a local proxy,

1

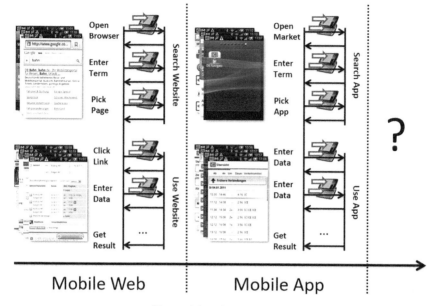

Figure 1.1   The Challenge.

the utilization of a more complex service, for example, to book a train ticket, requires users to specify numerous inputs such as destination, time, etc. using miniaturized and often, inadequate peripherals. As a consequence, the state of the art puts a natural limit on the complexity of the software and thus on the level of support that can be gained from existing services.

In contrast, ubiquitous computing [Wei91] envisions services which provide seamless and distraction-free support for simple and complex everyday tasks of their users. In order to realize this vision, the set of services available and the services themselves must be adapted to the users situation, behavior and to varying user intents. Thereby, adaptation must be performed autonomously in order to ensure that it does not conflict with the goal of providing a distraction-free user experience. This, in turn, requires services to gather a broad range of characteristics of the user's context at runtime. Examples for these characteristics include the user's location, activity, plans and goals.

Personal mobile devices such as smart mobile phones and personal digital assistants provide a promising basis for determining user context in an automated manner on a large scale. The reasons for this are manifold. First and foremost, personal mobile devices are self-contained and do not require additional infrastructure support, but existing cellular and wireless local area networks can provide the backbone for device interaction if needed. Secondly, though these devices are resource-constrained, newer generations are designed to support more complex tasks such as displaying a high-resolution movie. As a consequence, the devices are often not utilized to their fullest capacity, leaving enough resources to perform context recognition. Thirdly, with a variety of on-board sensor, personal mobile devices have access to both physical and virtual data sources, which allows multi-modal context recognition with high precision. Lastly, since the devices are carried by and owned by a single user continuously, the device's context is tightly correlated to the user's context and the recognition alone does not invade privacy.

In the past, these characteristics have contributed to the development of a number of context recognition systems for personal mobile devices. The recognition methods applied by existing systems are usually fine-tuned for specific requirements in order to provide reasonably accurate results while requiring limited resources. Although these methods are suitable for accurately detecting desired characteristics, they cover only a narrow set that can be detected by one device. Moreover, due to the resource-constrained nature of personal mobile devices, developers have usually concentrated on providing solutions for a concrete service.

The vision of ubiquitous computing, however, extends beyond the boundaries of a single service as it envisions seamless support for everyday tasks. As a consequence, achieving the overall vision of ubiquitous computing raises a number of challenges which include:

- the development of concepts to support the automated recognition of a broad range of context information types to support a variety of application scenarios in a generic fashion,
- the development of context recognition methods that are able to cope with the limited resource availability and energy constraints of personal mobile devices,
- the development of novel data acquisition and distribution protocols to share context information in order to increase the recognition accuracy without endangering privacy,
- the definition of an interoperable data representation model for context information and associated query models to support machine-to-machine communication,
- the design of a scalable data infrastructure to share and aggregate possibly frequently changing context information gathered by a large number of devices,

- the development of tools to reduce the required amount of manual configuration of policies and the mechanisms to validate them in order to protect the privacy of users,
- the design of new context-based human computer interaction techniques that are able to incorporate user goals and intents.

1.2 GAMBAS Objectives

The main objective of the GAMBAS project was to develop an innovative and adaptive data acquisition and processing middleware to enable the privacy-preserving and automated use of behavior-driven services that are able to adapt autonomously to the context of their users. Towards this end, GAMBAS was set up to address the complete set of challenges listed in the previous section in order to provide a truly integrated solution, thus closing a significant gap between the systems that were in use at the time and the vision of ubiquitous computing. The primary result of the project was the design, implementation and validation of a **G**eneric **A**daptive **M**iddleware, i.e. a set of application-independent services, to support the development and utilization of **B**ehavior-driven **A**utonomous **S**ervices.

As depicted in Figure 1.2, the GAMBAS middleware enables the development of novel applications and Internet-based services that utilize context information in order to adapt to the behavior of the user autonomously. To do this, the middleware provides the means to gather context in a generic, yet resource-efficient manner and it supports the privacy-preserving sharing of the acquired data. Thereby, it applies interoperable data representations which support scalable processing of data gathered from a large number of devices. In order to make the resulting services accessible to the user, the middleware supports intent-aware interaction, e.g., by providing recommendations for services, which minimizes the need for user inputs.

The realization of this middleware accompanied the development and integration of a flexible context recognition framework that is able to capture the context of users (e.g. location, activity, plans, intents), an interoperable data model to represent context information, a scalable data processing infrastructure to query and aggregate context information and to integrate context into services, a suite of security protocols to enforce the user's privacy when sharing context information and last but not least, a system to largely automate the discovery and selection of relevant services available to the user. In addition, it encompassed the development of tools to simplify the configuration of privacy policies, which ensures that the user's

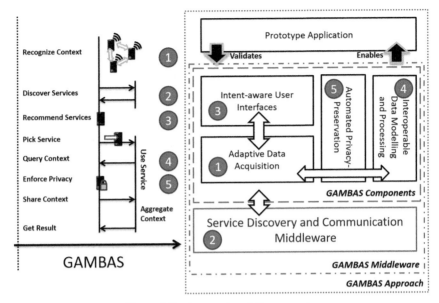

Figure 1.2 Approach and Components.

privacy expectations are met to improve the user experience and to increase user acceptance.

As a consequence, the implementation of this middleware by the members of the GAMBAS consortium resulted in a number of innovations in the research areas of context recognition, data modeling and processing and privacy preservation.

1.3 Application Scenarios

To define the scope of the vision addressed by the GAMBAS consortium, we first introduce the two scenarios that motivated the project. Thereafter, we discuss how they fit to the overall vision.

1.3.1 Mobility Scenario

John has just arrived to a new city. At the airport, he receives a message on his mobile phone by Bluetooth broadcast welcoming him to the city and inviting him to download an application on his smart phone in order to make his life in the city easier and to make his visit more enjoyable.

He follows the link proposed by the welcome message and downloads the application. When booting the application for the first time, he is requested to provide some data that does not affect his privacy. From that moment on, the application begins to capture the context related to John's interests including the change of positions, used transport modes, visited shops, etc.

The interface requests John to select which type of information he is interested in. John can choose from different sources of information and services. For this short visit to the city, John selects the mobility, events and shopping layers. The selection of "events" invites John to refine his selection and choose among different kinds of events: sports, theater, exhibitions, conferences, etc. John selects sports and theater.

The interface also suggests John to connect his smart phone application with social networks such as FourSquare, Facebook and Twitter. John selects Foursquare in order to publish his "check-in" events and share them with his friends in the city.

As it is the first time John visits the city, and he has just downloaded the application, the application is not able to predict the targeted destination of John. His city behavior profile has been just created and the phone's calendar is empty.

Thus, the application asks John: **What do you want to do?**

John responds via voice **I want to go to the hotel Astoria**. The smart interface of the application detects and recognizes the semantics of the phrase **go to** and **hotel Astoria** and suggests this destination. John confirms this selection with a simple gesture on his smart phone.

The application then shows John the route through public transport means to reach his destination. John begins his trip first by metro and then continues by bus. The application on his phone is able to detect at any time where John is and alerts John shortly before he has to leave the metro. Thereby, it notifies him about which bus to take next.

If he decides to leave the recommended route, he can do so at any point and at any time. If he decides to go for a walk in the city, he can leave the route and get updated route recommendations. At any point, he can look up information on the bus stops and metro stations or other points of interest (POI) making use of speech recognition combined with semantic services.

During the journey, the application informs John of the sport and theater events taking place in the next days in the city.

When he is close to his destination and since it is already lunch time, his smart phone suggests three restaurants nearby his hotel. At any point during his visit to the city, John can identify locations with a voice tag. At the location of the selected restaurant, he can use the voice recognition system to tag the location **Luigi's restaurant** or **good pasta**. Later on, the voice recognition will be able to use this information to lead John back to the restaurant.

After lunch, the application suggests buying in **Cortefiel** next to his hotel that has a two-for-one offer on spring shirts.

Once arrived at his destination, the application detects his **check-in** and suggests John to publish the event on his enabled social networks. John accepts the suggestion and according to his settings, his location is published in **Foursquare**. Once the goal is achieved, i.e. arrival at the destination – the application returns to its initial state, **What do you want to do?**

This time, John ignores his smart phone however. While John is in the city, the application keeps analyzing his behavior and suggesting information and services based on his position and preferences.

The application can notify John about shopping deals depending on his position and the proximity of the shops. Thus, the interaction with the user becomes more efficient and the GAMBAS framework is capable of filtering the offers, resulting in distraction-free support for the user's tasks.

1.3.2 Environmental Scenario

Paul is a regular user of the smart city application on his mobile phone. He uses it often to find the best options to get around in the city. For this, he is always subscribed to the mobility layer.

Today, he has decided to do some sport around the city, and his friend Ringo has explained him how to make use of the smart city application to obtain a jogging route through the less polluted areas of the city (CO_2 levels). He indicates the number of kilometers he wants to run, and for how long, and he also specifies that if possible he would like to run with a friend.

As a result, the smart city application offers him a route with Ringo. Paul observes that in order to have a reasonable route, the mobile application is proposing to take first a bus to the starting point of his jogging route.

At the same time, Ringo, who was already planning to go jogging, receives an alert asking him if he wants to share a route with Paul. He accepts and both friends receive a confirmation on the appointment in their agendas.

Ringo is not as concerned when it comes to environmental issues as Paul, so he does not use the public transport. Instead its smart city application proposes him a route by car through an urban tolling area. He is though quite concerned about costs, and by default he is subscribed to the mobility layer offering him a car pooling services. The urban tolling in the city depends on a number of factors such as type of vehicle used, number of passengers in the car and level of pollution in the city. Ringo receives a proposal from the application to share the trip with his friend George.

When activating the environmental layer on his mobile phone – in order to access the levels of CO_2 in the city – Paul has accepted to join the group of users collaborating with the municipality to study the noise levels in the city. Without any further intervention from his side, its mobile phone records and processes measurements of noise level each time Paul is outdoors and changes his position. At the end of the day, Paul can access the city pollution map application and check the noise levels in the route he has being following, including the jogging activity. Moreover, he obtains his environmental footprint due to the trip on public bus.

1.4 Overarching Vision

Given the advances in computer technology and the proliferation of wireless communication and sensing technologies, GAMBAS envisions the realization of major parts of the ubiquitous computing vision by means of a cloud of intelligent services, which provides adaptive and predictive information to people.

The basis for providing this information is the ability to automatically capture the state of the physical world by means of personal mobile devices as well as other sensing-enabled devices integrated in stationary or mobile Internet-connected objects. Given a variety of observations made by these sensors, the devices of a person can observe parts of its behavior which, in turn, can then be used to estimate and possibly predict parts of the person's behavior by means of a profile.

Upon request of the person, different views on this profile can be exposed (in a tightly controlled fashion) to different services such that they can adapt themselves not only to the person's current situation but also to some of the person's future intents. Thereby, the adaptive services might have to interact with other services as well as the personal mobile devices of other persons.

This creates dynamic mashups of services that share and integrate the information managed by them. To allow the ad hoc creation of such mashups, the information managed by each services and the information available on personal mobile devices must be discoverable. In addition, in order to seamlessly combine the information provided by different sources, it must be possible to easily link different pieces of information. This requires the use of a common, extensible and interoperable data representation to allow data processing that extends beyond the boundaries of a single service or device.

Based on the data provided by these dynamic service mashups, GAMBAS envisions new types of user interaction paradigms that transform the reactive information retrieval that is commonly applied by most Internet services into a proactive information provisioning that emerges from this system of Internet-connected objects and services.

1.4.1 Smart Cities

By employing the overarching vision described previously to the context of smart cities, it is possible to further detail the vision without narrowing its general applicability. GAMBAS envisions a smart city as a cloud of intelligent digital services that provides adaptive and predictive information to citizens. GAMBAS foresees a variety of services that manage different types of information that relates to the city as depicted in Figure 1.3.

Conceptually, these services and their data can be grouped into the so-called layers that cover different aspects of people's life in the city. A shopping layer, for example, might encompass services that manage store

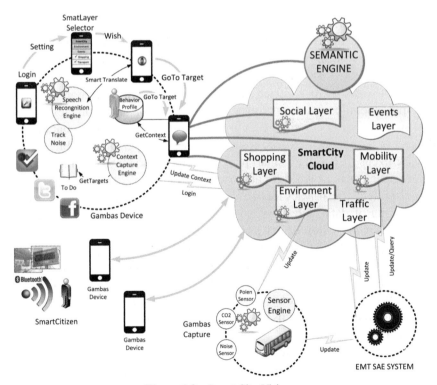

Figure 1.3 Smart City Vision.

locations and special offers or recommendations on products and experience reports on different stores. Similarly, a mobility layer might encompass services that manage taxi locations, bus routes, subway stations or traffic information. A social layer might manage relationships between citizens, events that take place in the city, bar and restaurant locations, recommendations, etc. An environmental layer might manage information related to water or air quality in the city or it might capture the noise levels at different places. Clearly, some of the services found in these layers can apply to multiple layers as some pieces of information and some of the services might be applicable to multiple aspects. As a simple example, both the shopping and the mobility layer may rely on generic geographic information about the smart city.

In order to enable the creation of dynamic mashups of services, the services export (parts of) their information. The information is then represented using an interoperable data representation that allows automatic linking of different pieces of information. This makes the information accessible to other services which can then add additional value by providing, for example, a better experience for a specific group of citizens. In order to simplify the integration of services, a distributed query processing system enables the execution of queries across different information sources.

To provide up-to-date information and adaptive information to users, the layers capture information from different sensors embedded in various Internet-connected objects. The objects may belong either to a particular service provider or to a citizen. The devices in the first category, may, for example, encompass sensors embedded in a taxi or a bus or they may be deployed at specific positions such as a bus stop or a metro station. The devices in the second category may encompass the personal mobile devices of the citizens such as their smart phones but they also may contain traditional systems such as their desktops at home, for example.

To protect the privacy of the citizens, they can control the collection and sharing of data with the services in different layers. Towards this end, behavioral data is stored and processed on the devices that belong to the citizen. Optionally, in order to access additional services, they may share their information with specific service providers or other citizens. In order to avoid the expensive task of manually controlling the sharing process, automatic proposals for different settings can be computed based on social relationships that are formalized by means of existing policies that the citizens created for similar contexts.

To access the information from services and to perceive their current context, citizens will run a special application on their personal devices. The application performs predictions based on the citizens' past behavior. Given these predictions, the application is able to proactively retrieve information from different layers that are interesting for a particular citizen. Furthermore, it automatically determines appropriate times to notify the citizen about important events. For example, when traveling in a bus, the application may notify the citizen shortly before the bus stops at the target destination such that the citizen does not miss the bus stop. In cases where the citizen is exploring new terrains that cannot be predicted, a natural user interface based on speech recognition technology allows the citizen to specify alternative goals.

1.4.2 Characteristics

Based on this smart city architecture, it is possible to identify two key characteristics that differentiate the basic idea from other approaches in this application domain:

- **Adaptive Acquisition and Presentation:** In terms of data acquisition, the GAMBAS vision foresees citizens not only as consumers of digital services, but also as an important source of information that can provide feedback to different stakeholders. This feedback can then be used to adapt services, which results in mutual benefits for both the citizens and the providers of services.
- **Dynamic and Distributed Processing:** In terms of data processing, the GAMBAS vision foresees high dynamics that depend on the individual behavior of citizens as well as the results of their aggregation. This enables novel services that go beyond the possibilities of today's service infrastructures as they are typically focusing on isolated operation (often referred to as "data silos") or they solely combine a few data sources that are determined statically.

1.5 State of the Art

Although it has not been realized so far, overall, the vision of ubiquitous computing as defined by Mark Weiser [Wei91] is not new. Ever since its formulation in 1991, researchers and practitioners have focused on closing different research gaps. With respect to middleware issues, a significant amount of research has been performed in the area of enabling seamless device interaction as well as application adaptation. Furthermore, there have been considerable efforts in the area of enabling context management, which is an important basis for context-adaptive applications. Lately, the availability of results in these areas has led to the development of a number of large-scale sensing applications. In the following, we briefly review the state of the art in each of these fields, but before we do this, we quickly review the available hardware technologies.

1.5.1 Hardware Technologies

As for any other software project, the execution environment for the GAMBAS middleware is defined by a subset of the existing hardware platforms. Due to the specific focus of GAMBAS on enabling adaptive data acquisition with Internet-connected objects, in the following, we briefly discuss the available hardware technologies with respect to devices, communication and sensing. The focus is not to provide a comprehensive list of available technologies. Instead, we take a more high-level perspective that uses current technology as examples but, in principle, is independent of the concrete implementation. Based on the resulting discussion of device, communication and sensing technology, we then introduce a classification of device types that are the basis for GAMBAS. Thereby, it is noteworthy to mention that not all features of the GAMBAS middleware are realized for all types of devices. However, it enables their integration into a single system.

1.5.1.1 Devices

The devices forming the Internet of Things are heterogeneous. For example, besides traditional personal computer systems, a significant number of devices are either mobile or integrated. When analyzing the different types of devices, we can categorize them with respect to several orthogonal axes.

1. **Specialization:** Naturally, we can classify devices on the degree of specialization. This degree may range from general purpose devices such as PCs or laptops to special purpose devices such as micro-controllers that are integrated into all kinds of objects. Although, in principle, the concepts developed by GAMBAS are applicable to all kinds of Internet-connected objects, the GAMBAS middleware does not focus on the latter. The reason for this is that highly specialized devices are often closed systems that cannot be programmed easily. However, given the rapid advances of technology, we can expect that many closed devices will open up in the future.

2. **Resources:** Independent of the degree of specialization, we can classify devices on the available resources. On the one end of the spectrum, the set of devices forming the Internet of Things may encompass resource-rich devices such as mainframes or clusters of workstations. On the other end, they may contain resource-poor devices such as simple sensor

nodes. In between, there are devices such as laptops or devices with less resources such as mobile phones or tablets.

3. **Mobility:** Another important axis is the degree of mobility. Here, we can distinguish stationary devices and mobile devices. In contrast to stationary devices, mobile devices are usually equipped with batteries and thus, their energy is a limited resource that needs to be managed appropriately. This is especially true, when using mobile devices for long-running tasks such as the continuous monitoring of the environment.

4. **Interaction:** Last but not least, the devices can also be classified based on their capability of supporting immediate interaction with a user. Here, we can distinguish devices that support user inputs, e.g., by means of graphical or audible interfaces, and devices that are invisibly integrated other objects. This axis is particularly relevant since only devices that support the interaction with a user can be configured manually by the user. Due to the invisible integration, the remaining devices can solely be configured indirectly through other devices.

1.5.1.2 Communication

Existing communication technologies can be broadly classified into wired and wireless. Due to the success of mobile devices, the latter ones have become main stream over the last couple of years. At the present time, there are several technologies that are widely available and frequently integrated into mobile devices. They cover the complete spectrum from low to high speeds and low to high range. At the same time, they exhibit vastly different energy profiles.

- **Near-Field Communication** is a set of standards to enable radio communication between devices by bringing them in close proximity. NFC is based on existing standards on radio frequency identification (RFID). In contrast to other technologies in that family, it enables bi-directional communication between two devices. However, it offers only low transmission speeds and it is only applicable to very close range communication (i.e. few centimeters). At the present time, it is mostly used for mobile payment systems or in order to bootstrap connections with other communication technologies (e.g., Bluetooth).

- **ZigBee** is a standard for short-range communication. ZigBee is specifically designed for low-power devices with low data rate and short-range communication capabilities. The IEEE standard 802.15.4 defines the physical and the MAC layer for ZigBee. The devices in a ZigBee setup can be categorized into ZigBee coordinators, ZigBee routers and ZigBee devices. The ZigBee coordinator is the central entity that keeps record of the devices in the network as well of the other ZigBee coordinators. The ZigBee router is responsible for routing messages and associating devices with each other. Devices that are not ZigBee coordinators or ZigBee routers are classified as ZigBee devices.

- **Bluetooth** is another popular short-range communication standard. Bluetooth modules are commonly available for standard computers and various peripherals. These modules support low-bandwidth and short-range communication. Depending on the communication range and energy consumption, Bluetooth devices are divided into three classes. Class 1 Bluetooth devices consume around 100 mW and support approximately 100 m. Class 2 Bluetooth devices consumes up to 2.5 mW and support communication range of approximately 10 m. Class 3 consumes the minimal power (1 mW) but also provide the shortest communication range (approximately 1 m).

- **Wi-Fi** is probably the most popular communication standard for connecting various devices such as laptops or mobile phones wirelessly. Wi-Fi certification is given to the devices with wireless capabilities that implement IEEE 802.11 standards. There exist several 802.11 standards that include 802.11a, 802.11b, 802.11g and 802.11n. Wi-Fi-supported routers cover approximately 100 m in outdoors. Since the clients in the Wi-Fi network do not require wire, the network can be easily extended. Wi-Fi-enabled devices can move in a limited area but they have relatively short range. A possible shortcoming of Wi-Fi-certified devices is that they have comparatively higher energy requirements.

- **UMTS** (Universal Mobile Telecommunications System) is the successor of GSM and designed to support third-generation telephone technology. UMTS is specifically designed to support advanced services. It is developed to support 14 Mbps data transfer rate and UMTS support is now commonly available in most smart phones. Compared to its predecessor (GSM), it consumes more power. However, in terms of speed and service capabilities, it is a significant improvement over GSM.

For stationary devices, wired communication technologies are still an important alternative to wireless technology. Many stationary general-purpose devices such as servers are usually connected with Ethernet.

- **Ethernet** is based on IEEE 802.3 specification and is a very popular LAN technology. The specification defines standard for physical layer as well as data link layer of the OSI model. Starting with 10 Mbps, it has evolved to support 100 Mbps (fast Ethernet) and later 1000 Mbps

(Gigabit Ethernet) speed. Currently, the fastest speed standard supported by Ethernet is 10 Gbps, although we can assume that there will be further progress on connection speeds.

1.5.1.3 Sensing

Besides device and communication technologies, the third hardware pillar of GAMBAS is sensing technology. Over the last couple of years, device manufacturers have started to integrate various sensors into different types of devices. At the present time, current mobile devices such as smart phones and tablets commonly exhibit the following combination of sensors:

- **Accelerometer:** Accelerometers are used to measure the acceleration that a device experiences. In most cases, they are able to differentiate acceleration along three axes. Usually, they are used to adapt the screen orientation of the device according to the way the user is holding it. However, researchers have also used accelerometers for various other tasks such as classifying the mode of locomotion or detecting potholes.
- **Gyroscope:** More recently, device manufacturers have started to add gyroscopes to the set of standard sensors that are available on mobile phones. Gyroscopes are used to measure the orientation of a device. Advanced applications include inertial navigation systems, for example. However, at the present time, they are mostly used to support gaming.
- **Microphone:** As a natural consequence of their function, all mobile phones are equipped with microphones that allow them to record and transmit voice during a call. However, in addition to that, most devices nowadays exhibit multiple microphones (e.g. to enable automatic noise reduction) that can also be used to capture and analyze ambient sound.
- **Proximity:** In order to activate and deactivate the screen automatically during a call, many mobile phones are equipped with proximity sensors that can measure the distance between the phone and another object (typically in front of the screen) in a course-grained scale (e.g. far or close).
- **GPS:** To support location-based services and to support user navigation, many mobile devices are equipped with GPS receivers. Although they cannot be used reliably in indoor environments, outdoors they provide reliable localization with 5 to 10 m accuracy.
- **Camera:** Similar to microphones, nowadays, most mobile phones and tablets are equipped with cameras which can be used to record videos as well as still images. In addition to simply taking pictures or recording videos, they can also be used to recognize visual tags (e.g. QR-Tags) and

they can be used for different types of context recognition applications (e.g. to automatically detect gas station prices).

- **RF:** Although they are mostly intended for communication, RF-based communication technologies such as Wi-Fi or GSM can also be used to extend the capabilities of other sensors such as GPS, for example. The use of these technologies as sensors usually requires special maps that model the signal propagation in a certain area. Using these maps, a course-grained but energy-efficient localization can be supported.

Besides mobile devices, researchers have also developed a number of sensing platforms such as Berkeley Mica2 or UCLA iBadge, etc., mostly in the area of wireless sensor networks. Typically, these platforms can be extended with different types of sensors, but most of them contain at least the following combination of sensors.

- **Light:** Light sensors typically measure the light level received at a particular point of the device. In many cases, light sensors are directly built into the sensor node or they can be added by attaching a sensor board.
- **Temperature:** Temperature sensors typically measure the ambient temperature of the sensor node. In many cases, the sensors are not calibrated and the raw values need to be converted programmatically to the usual Celsius or Fahrenheit scale.
- **Pressure:** Pressure sensors typically measure the barometric pressure, and thus, they can be used to compute the altitude.
- **Humidity:** Humidity sensors typically measure the humidity using capacitive measurements. In many cases, they are bundled with temperature sensors.

In addition to mobile devices and sensor nodes, there are numerous application-specific sensors. Due to their great variety, it is not possible to provide a comprehensive list here. To name some examples that may be relevant in the context of GAMBAS, using the OBD unit of a modern car, it is possible to capture various engine-related values. These include, for example, the current fuel consumption or the current state of the catalytic converter.

1.5.1.4 Classification

Based on the previous discussion of device, communication and sensing technologies, we can identify four broad classes of devices that are forming the hardware platform for services developed with the GAMBAS middleware.

Intuitively, based on the capabilities of the device, the support provided by the middleware differs.

- **Back-end computer system (BCS):** Back-end systems usually consist of one (or more) general-purpose computer that is connected to the Internet via a wired and often high-speed connection. Usually, they exhibit high storage and processing capacities and they are shared by multiple users remotely, i.e. through the Internet. Consequently, to most of their users, they do not expose a physical interface that would enable interaction. Instead, they are accessed through web-browsers or via custom applications that are performing some form of remote call (e.g. RPC, RMI, etc.). The GAMBAS middleware uses these systems for data storage, aggregation and processing.

- **Traditional computer system (TCS):** Traditional computer systems encompass workstations, desktops and laptops. If they are stationary, they are typically connected via wired connections. If they are mobile, like laptops, the predominant communication technology is Wi-Fi. In some cases, they are equipped with a few sensors (e.g. microphones, cameras, accelerometers for hard disk protection). Usually, they are accessed and used by a single user (e.g. personal desktop/laptop) or a small group (e.g. shared workstation). Although, they have fewer resources than most back-end computer systems, when considering that they are not shared between many users, the ratio of resources to number of users may be equally high. The GAMBAS middleware uses these systems primarily to perform similar tasks as back-end systems (although on a smaller scale). However, it also enables their usage as sensing devices.

- **Constrained computer system (CCS):** Constrained computer systems include mobile devices such as smart phones and tablets. Furthermore, they include stationary devices such as set top boxes or industrial PCs. When compared with traditional computer systems, they exhibit a significantly lower amount of computing resources with less capable processor architectures (e.g. ARM instead of X64) and less amount of memory (e.g. MB instead of GB). In many cases, they are equipped with a multitude of built-in sensors (e.g. mobile phone) or they can be attached to application-specific sensors (e.g. industrial PC). Consequently, they provide the primary basis for data acquisition in GAMBAS. In addition, they are also used as a personal data storage that can be accessed remotely.

- **Embedded computer system (ECS):** Embedded computer systems include highly specialized micro-controllers or ASICs that are built into existing products such as a dishwasher or a car. Furthermore, they include less specialized sensor platforms that may be programmable such as a SunSPOT or a Mica2 node. Usually, these systems are not directly connected to the Internet. Instead, they can be connected through some gateway device that mediates the interaction. Although such embedded devices outnumber the other classes, the GAMBAS middleware does not focus on the use of these devices as a primary processing platform. The reason for this is that usually, these devices cannot provide their function without additional computing infrastructure. Furthermore, in many cases, they are not equipped with easily accessible interfaces or they do not provide sufficient computing resources to implement additional services. However, the GAMBAS middleware supports their use as data sources when combined with a more capable device such as a constrained computer system or a traditional computer system.

1.5.2 Communication Middleware

Regarding device interaction, researchers and practitioners have developed a number of communication middleware systems to enable the seamless and trustworthy cooperation of a heterogeneous set of possibly resource-poor connected objects. Examples for past and present research projects in this area are the 3PC [3PC12], GAIA [RJH02] and AURA [GSSS02] projects or the PECES FP7 project [PEC12], to name a few. Traditionally, the resulting systems either focused on enabling the interaction of devices at a specific geographic location (i.e. the so-called smart spaces) or focused on enabling the interaction of devices that are in close proximity (i.e. the so-called smart peer groups). More recently, systems such as the PECES middleware have integrated and extended these two concepts by enabling the interaction within a smart space that is formed by devices in close proximity and beyond smart spaces by enabling device interaction across the Internet in a peer-to-peer fashion. Since the resulting concepts provide a higher degree of flexibility, the GAMBAS middleware will use PECES as its underlying communication middleware.

Besides device interaction, research on communication middleware also addressed the development of new programming paradigms to support the development of adaptive applications, for example, on the basis of goals as done by O2S [PPS$^+$08] or components as done by PCOM [Han09] or flows as done in the ALLOW FP7 project [ALL12]. While these abstractions are interesting to support the development of adaptive applications, the GAMBAS project does not primarily target the development of new abstractions to support application adaptation. Instead, it focuses on the acquisition of context information as well as the processing of environmental information in a privacy-preserving way, which usually provides an important basis for adaptation that is independent of the concrete abstraction that performs the adaptation. Consequently, from a high-level perspective, the goal of GAMBAS is a more fundamental one that will enable the use of such abstractions at a later point in the development process.

1.5.3 Context Management

The importance of context information for the realization of ubiquitous computing has been recognized very early after the formulation of the vision [SAW94]. Over the course of several years, researchers have developed a number of middleware systems to acquire and leverage context information, e.g. [HKL$^+$99], [SDA99], [Bar05]. Traditionally, these systems have either focused on the scalability issues that arise from providing context awareness in an application-independent way using a federated system [HKL$^+$99] or focused on the actual distributed acquisition and usage when developing applications with a limited scale such as a room or a house [SDA99], [Bar05]. In addition to that, specialized context management systems have been integrated into all kinds of middleware systems for smart environments such as GAIA [RJH02] and AURA [GSSS02], to name a few. Similar to [SDA99] and [Bar05], these systems focused on a rather restricted execution environment.

Besides that, the active research in the area of sensor networks and cooperating objects has spawned a number of initiatives to acquire context information from a heterogeneous set of networked sensors that is deployed in an environment. Project at the European level include, for example, the PLANET FP7 project [PLA12] which works on concepts to deploy and operate large-scale sensor networks to capture environmental information. However, usually these systems focus on low-level networking aspects of various sensors or they solve high-level data management aspects resulting from a large number of sensors. Thereby, these systems do not have to consider the resulting privacy implications when moving from environmental context – such as temperature or animal population – to personal context – such as human location, activity and plans.

1.5.4 Sensing Applications

In the recent past, the advances with respect to middleware, device and sensing technologies have led to the development of a number of large-scale sensing applications that are often summarized as participatory sensing [BEH+06] or people-centric sensing [CEL+06] applications. Similar to the goals of GAMBAS, these applications leverage the personal Internet-connected objects of users to capture relevant sensor information. The type of information typically depends heavily on the application area. To give some examples, DietSense [RSB+09] tries to collect diet-related information about the user through photos and sound samples. PEIR [MRS+09] provides an estimate of the environmental impact of a user trip by determining the mode of locomotion. BikeNet [EML+10] captures the biking experience by means of measuring the location and speed and providing an estimate over the used calories. Haze Watch [CYCS12] captures pollution information by attaching external sensors to a mobile phone.

Usually, these and other similar types of applications capture the sensor information at some central application server where it is then processed and analyzed. Furthermore, although they are very similar, they are often built completely from scratch without adequate middleware support. Finally, in most cases, the applications merely inform the user about the collected data by providing some aggregated view on it. The GAMBAS middleware simplifies the development of such applications by providing a scalable, interoperable basis. In contrast to collecting all data at some trustworthy central server, however, the GAMBAS middleware provides configurable sharing that enables users to protect their privacy, if that is desired, which allows users to balance the potential loss of privacy with the potential gaining in service quality. Furthermore, instead of merely aggregating and visualizing the information, the middleware enables the behavior-driven adaptation of services.

1.6 Innovations

Building upon the existing work, the GAMBAS middleware specifically targets the acquisition of personal context information. Consequently, it shares similar goals with several of the existing large-scale sensing applications. However, in contrast to existing applications, GAMBAS also can enforce the user's privacy goals. Towards this end, the acquisition is performed primarily with personal Internet-connected objects. This empowers the user to limit the sharing of the acquired context. In order not to overwhelm the user, the GAMBAS middleware contains a framework to automate the sharing in a privacy-preserving manner. Furthermore, to directly use the acquired context on the connected object, the middleware provides concepts to implement intent-aware user interfaces, which allows the user to have full control over the use of the GAMBAS software via a fine granular system to enable and disable features as needed. Finally, in order to use the shared context effectively in enterprise business processes, the middleware makes use of an interoperable data representation with the associated processing infrastructure that supports a large number of sensors. This provides the basis for efficient object–object interactions and thus, it enables the development of services that can autonomously adapt to the user's behavior.

2

Architecture

This chapter describes the high-level architecture of the GAMBAS middleware. To clarify the architecture, the chapter first presents a static perspective that focuses on the identification and definition of entities that are operating different parts of the architecture (operational view), building blocks that constitute the architecture (component view) as well as types of information that are handled by the architecture (data view). After presenting the static perspective on the architecture, the chapter introduces a dynamic perspective that focuses on a description of the interaction of architectural components. To do this, the dynamic perspective provides details on the acquisition of data (acquisition view), the discovery of data and the respective processing of queries (processing view) and the usage of data for inferences (inference view). Finally, to clarify the interactions, the chapter discusses the interfaces between the different components.

2.1 Static Perspective

The static perspective introduces the entities that interact with each other in order to produce and consume services. Furthermore, it introduces a functional breakdown into a number of core building blocks. Finally, it discusses different characteristics of the data that shall be handled by these building blocks in order to facilitate the envisioned creation and usage of behavior-based autonomous services. The dynamic perspective, which is discussed later on, ties these entities and building blocks together by describing how the different types of data are exchanged among the building blocks in order to achieve different goals of entities.

2.1.1 Operational View

As basis for the further discussion of the functional building blocks, it is important to clarify the roles of different parties that may be involved in the operation of various parts of the architecture. It is worth mentioning that a single entity may exhibit a number of roles simultaneously and the roles that it adopts may also depend on the specific type of data that might be processed by a service. Furthermore, it is also possible to look at the infrastructure from different angles. For example, we might take a data-oriented view and classify the entities as either data acquirers or data aggregators. In the following, however, we look at the operation of the infrastructure from a service-centric perspective, as this clarifies the entities and also highlights the innovative features that are targeted by the project.

As depicted in Figure 2.1, from a high-level service-centric perspective, the entities involved in the GAMBAS architecture can exhibit one of more of the following three roles:

- **Service operators:** A service operator is responsible for executing and maintaining a part of the software and hardware infrastructure that is required for a particular service or a set of services. The operator provides computing resources such as processing capabilities and storage capacities in such a way that they can be accessed remotely

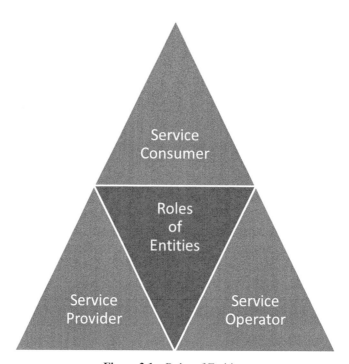

Figure 2.1 Roles of Entities.

via the network. The operator is not responsible for the actual functionality provided by the service. Instead, the service operator simply provides the basic infrastructure, possibly with service-level agreements on the performance of the system. To protect the privacy of sensitive data, we envision that some generic services such as the distributed stream processing between employees of the same enterprise or between groups of friends might be operated by an enterprise or by one or more citizens. To ease the provisioning of such generic services, we envision that service operators might offer them as pre-installed and managed bundles, similar to existing website or cloud computing offerings. Alternatively, technically advanced users might run the necessary software components on their own Internet-connected home server or wireless home router.

- **Service providers:** While a service operator is solely providing thetechnical basis for the execution of a particular service, the service provider is the actual responsible entity that offers the service to different users. In many cases, the service provider will be interested in offering a particular set of data to the consumers of the service. Furthermore, the service provider might also be interested in collecting some information from the service consumers which can then be used, for example, to improve the quality of the service. In this case, the service provider actually becomes the consumer (of parts) of the provided service. Besides service providers that want to share a particular data set, we also envision service providers that simply combine existing data sets (possibly offered by different service providers) in order to add value. For example, one service provider might combine the social graph of a set of persons with their travel behavior in order to provide recommendations for trip routes. To do this, the service provider might have to access the social graph and the trip routes from two services that are controlled by another provider. To enable such service mashups, the GAMBAS middleware uses an interoperable data representation that is based on linked open data principles.

- **Service consumers:** The last role foreseen by the architecture is the actual consumer of a service. The service consumer accesses the data and functionality offered by a service provider using the infrastructure of the service operator in order to ease their everyday tasks. In many cases, the consumer will be an end user that is accessing a particular service from a mobile Internet-connected device. Thereby, the end user might not have to initiate the interaction. Instead, the intent-aware user interface might initiate interactions via the middleware at the right point in time without manual intervention. In order to improve the service quality available to them, the end users might be willing to opt-in to collect additional data that can help the service provider to improve the service. This mutually beneficial data collection and sharing forms the basis for the second group of service consumers. In addition to the end users, we envision that service providers can become the consumers of their own services. As a simple example, consider that a city might offer a service that enables users to share and report air quality measurements. The end users, i.e., the citizens, might then use the resulting air quality map to avoid polluted parts of the city. In addition to this, the service provider, i.e. the municipality, might use the service in order to dynamically adjust the road toll on different streets of the city in order to improve the minimum air quality.

Although these three roles are not new and are at the core of most service-oriented infrastructures, their interpretation in the context of the GAMBAS project is more dynamic. The basis for this is formed by the two key characteristics of the overall vision, namely the adaptive data acquisition and presentation as well as the dynamic and distributed data processing. Due to the former, the service consumers may also contribute to the provisioning of a service by collecting and sharing some data using their Internet-connected mobile devices. This, in turn, can result in mutual benefits for the service providers and the consumers. For some services, the providers may become the consumers of their own services. Due to the latter, new types of service providers may emerge in the network. Instead of providing their own data sets, they may simply link the existing data sets in novel ways – possibly enriching them with additional data. This will allow more tailored and specialized services and it should lead to a more thorough support for various types of service consumers that may exhibit different behaviors.

2.1.2 Component View

Intuitively, as a service-oriented architecture that is supposed to be capturing and delivering data, we can identify three main building blocks which provide data acquisition, data storage and distributed data processing. On top of that, in order to enable the limited sharing of data, we can furthermore identify a building block that is responsible for managing the data access. Finally, in order to remotely retrieve the necessary data in an automatic fashion, we can identify a building block that takes care of data presentation. In the following, we describe these building blocks in more detail:

- **Data Acquisition Framework (DQF):** A primary capability of the GAMBAS middleware is its ability to automatically capture data on behalf of the end user or a service provider. For this, the middleware encompasses a data acquisition framework that is capable of running on different types of devices. Based on the four device classes introduced in Section 1.5.1.4, the data acquisition framework primarily targets Constrained Computer Systems (CCS). In addition to that, the data acquisition framework provides support for Embedded Computer Systems (ECS) by means of connecting the embedded systems to a constrained system. The data acquisition framework provides generic and extensible support for virtual and physical sensors, and it optimizes the data acquisition with respect to energy consumption. Furthermore, in order to support applications, the data acquisition framework provides a number of example activities and intent recognition components that primarily deal with location information, movement modalities, bus routes and environmental information. The data that is captured using the acquisition framework can either be stored locally on the device or it can be forwarded automatically to a particular service that is connected to the Internet. The former approach can be taken in order to protect the user's privacy when dealing with privacy sensitive data, whereas the later approach can be taken with data that does not impact the user's privacy or which is explicitly shared on behalf of the user.

- **Semantic Data Storage (SDS):** To store the data of the user on a local device or at a particular service, the architecture introduces a semantic data storage component. Similar to the data acquisition framework, the semantic data storage is primarily targeted at device classes with more resources, such as Constrained (CCS), Traditional (TCS) and Backend Computer Systems (BCS). The data that is stored in a semantic data storage component follows the linked open data principles and uses the interoperable data representations that have been developed as part of the GAMBAS middleware. Furthermore, the data storage is able to interface with different types of query processors, depending on the resources available on the device. This implies that there may be different implementations of this component that are optimized for different device classes.

- **Legacy Data Wrapper (LDW):** The semantic data storage component is primarily targeted at the management of interoperable data that is following the linked open data principles. However, in the short term to mid-term, it is unrealistic to expect that all types of information that are interesting for a service consumer are modeled with this approach. Consequently, it is necessary to integrate with data that is stored in an existing "data silo" using a proprietary data representation. In order to smoothen the transition from proprietary to interoperable representations, the architecture explicitly foresees legacy data wrapper components that transform the data and possible functionality provided by legacy services into an interoperable data representation. Intuitively, it is not possible to provide a generic legacy data wrapper that can handle all possible data representations. Instead, the GAMBAS middleware encompasses basic software that eases the development of an application-specific data wrapper. Thereby, the basic software primarily targets Traditional (TCS) and Backend Computer Systems (BCS) as these are commonly used to manage data and to provide services.

- **Query Processors (xQP):** In order to make the data stored in semantic data storages available to services and applications, the architecture introduces query processor components that are capable of executing queries on top of the storages. As described in detail in Chapter 4, the query language that is supported by the query processors is a subset of the SPARQL language that considers the limited resources available on Constrained Computer Systems (CCS). Due to the different dynamics of different types of data that is handled by the GAMBAS middleware and due to the different amounts of resources that are available on different classes of devices, the architecture divides the query processors into the following two components:

  ○ **One-time Query Processor (OQP):** The one-time query processor is targeted at the execution of queries that evaluate the current state of the data in the semantic data storages. It executes queries that produce a single result based on the current information and the specific query. Consequently, this query processor is targeted at static information that does not change frequently or at applications that only require a one-time view. From a resource perspective, the one-time query processor is designed to support a broad range of devices including Constrained (CCS), Traditional (TCS) and Backend Computer Systems (BCS). Due to resource constrains, one-time query processors in CCS are limited to process only data stored in semantic data storages belonging to the same system. OQPs in less constrained devices have access to remote semantic data storages to allow the combination of data from multiple sources. As explained later on, the provisioning remote access respects the privacy constrains.

○ **Continuous Query Processor (CQP):** In contrast to the one-time query processor, the continuous query processor is specifically targeted at dynamic data. It executes queries that can produce multiple results based on the changes to the underlying information and the specific query. Consequently, this query processor is suitable for services and applications that require continuous monitoring of events that might be captured by multiple data sources. However, in order to handle such queries, it is necessary to introduce buffers that can easily exceed the resources available on Constrained Computer Systems (CCS). Consequently, this type of query processor will be targeted at Traditional Computer Systems (TCS) and Backend Computer Systems (BCS). Yet, in order to evaluate continuous queries, Constrained Computer Systems may make use of continuous query processors that are provided as a service that is operated by a third party. Towards this end, the continuous query processor can be considered to be a generic component that can be deployed by different entities, provided that they have access to a suitable Internet-connected computer system. Similar to OQPs, access to remote data is also enabled in continuous query processors.

- **Data Discovery Registry (DDR):** To enable transparent distributed query processing, the query processors must be able to discover the data sources that are available on the network. To make the data discoverable, a device may announce the data available in the semantic data storage to the data discovery registry which in turn will typically use a semantic data storage component to manage the announcements. In case of personal mobile devices, the announcement may be limited or modified depending on the privacy preferences of a particular end user. To enable this, the semantic data storage and the data discovery registry must interface with the privacy framework.

- **Privacy Framework (PRF):** Given the above components, it is possible to acquire information using all types of systems. Furthermore, it is possible to access dynamic as well as static information using one-time and continuous queries. In principle, this is sufficient to enable the acquisition and sharing of data. However, as some data such as the end user location or the end user travel preferences might be sensitive from a privacy perspective, it is necessary to limit the data acquisition and in particular the data sharing such that it respects the privacy preferences of different entities. Achieving this is the primary task of the privacy framework. Conceptually, the framework interacts with the semantic data storage as well as the data acquisition framework that is deployed on each personal device. In addition, the privacy framework may also be used to limit the access to information that is provided by a particular service. For this, it is integrated into the device that is offering the service.

Using a privacy policy that can be generated automatically by means of plug-ins that access proprietary data sources, the privacy framework takes care of exporting sensitive data in such a way that it can only be accessed by legitimate entities. Furthermore, depending on the user preferences, it can apply obfuscation in order to limit the data precision and it may anonymize the data in order to unlink the data from a particular user. Since the GAMBAS middleware targets the use of personal mobile devices as primary sources of data, the privacy framework not only supports Traditional Computer Systems (TCS) but also Constrained Computer Systems (CCS) as its execution platform.

- **Intent-aware User Interface (IUI):** As the last building block of the architecture, the intent-aware user interface is responsible for leveraging the remaining components in such a way that the end user ideally receives the right information at the right time. To do this, the intent-aware user interface executes queries against different services based on the behavior of the user and decides on how and when to present what information to the user. Since the past behavior of the user might not be sufficient to predict new user goals, the intent-aware user interface can also provide ways of allowing the user to modify the predicted behavior. Furthermore, as it is the primary component that is visible to the user, it has to support manual customization by the user. This encompasses, for example, the selection of layers that are interesting for a user or the manual tweaking of a generated privacy policy in a user-friendly way. Although we envision that the concepts behind the intent-aware user interface are applicable to different types of devices, we assume that in the short term and mid-term, they will be most useful for users when they are presented on their personal mobile devices. Consequently, the current implementation of the GAMBAS middleware focuses primarily on Constrained Computer Systems (CCS).

2.1.3 Data View

The GAMBAS architecture aims at supporting a broad range of services and applications whose data exhibits vastly different characteristics.

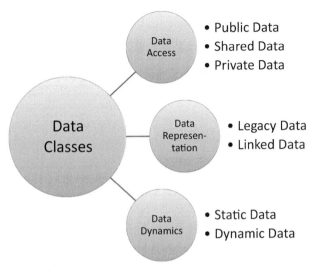

Figure 2.2 Classes of Data.

Depending on the point of view, it is possible to classify these characteristics along various orthogonal dimensions. As depicted in Figure 2.2, we focus on the level of data access, the type of data representation and the dynamics of the data. In the following, we take a closer look at these three dimensions and describe how the different classes of data are handled by the GAMBAS middleware architecture.

2.1.3.1 Data Access

Given the fact that GAMBAS aims at supporting the development of behavior-driven services that adapt autonomously to the user, it is clear that the GAMBAS architecture must be able to thoroughly support different levels of access to data, especially in cases where the collected data may be sensitive in terms of privacy. Based on the level of access, we can identify the following categories:

- **Public Data:** Public data may belong to an individual or an organization which makes the data available to third parties. Thereby, the entity that owns the data grants free access to all data for all other entities. Examples of such data could be stock prices, weather information, etc. We can assume that many applications will require public data to provide relevant and useful services. Although we can assume that most public data will be provided by services that are executed on resource-rich devices connected continuously to the Internet, the GAMBAS architecture also allows the provisioning of public data by means of Constrained Computer Systems (CCS) such as a mobile phone. To enable seamless discovery of public data, however, the device responsible for the data must publish the metadata in the Data Discovery Registry (DDR).

- **Private Data:** In most cases, private data belongs to an individual person and it could be the user's personal data or data that the user is not willing to share with everyone. Examples of such data could be the user's contact information or the user's current location. In addition, private data may also reflect the internal data of an enterprise that is not supposed to be shared with other entities. For this type of data, the GAMBAS consortium made the deliberate decision to limit its distribution. Although it might be more practical to provide online access to private data, the GAMBAS architecture foresees the storage of private data exclusively on the devices that own it in order to prevent illegitimate access and processing through third parties. Consequently, the private data will remain on the devices that collected it unless the responsible entity makes a deliberate decision to share (parts of) it.

- **Shared Data:** In many cases, limiting the types of data to only private and public can be overly constraining. Depending on the user's preferences or on the business model of an enterprise, it might be more beneficial to share (parts of) the private information in order to get better services or to increase the revenue. For both cases, the GAMBAS architecture foresees support for shared data. In essence, shared data is a particular view on the private data. This view can be accessed by other entities that are authorized. In order to safely support shared data, it is necessary to enable trustworthy authentication among the different entities and there needs to be a policy that details who will gain access to which view on the data. Managing this process and the associated policies is done by means of the privacy framework that is an integral part of the architecture. The privacy framework thereby ensures that only legitimate entities will be able to access a shared view.

2.1.3.2 Data Representation

As hinted in the component view, in the short term and mid-term, we cannot assume that all types of data will be represented using the models and approaches developed by the GAMBAS project. Instead, we must ease the integration of existing data that may be represented using proprietary formats by means of legacy data wrappers. Consequently, based on the level of integration, we can identify the following two classes of data:

- **Linked Data:** Linked data represents data that follows the linked open data principles that are the basis for the interoperable data representation used by the GAMBAS middleware. Using the semantic data storage component, it is possible to store linked data on any supported device. Furthermore, using the one-time and the continuous query processor, it is possible to query the static and dynamic data stored in one or more semantic data storage components. To implement interoperable services and to ensure that it is possible to easily create composed services, it is necessary that the information is represented using this data representation.

- **Legacy Data:** Although there are good reasons for picking up the interperable data representations promoted by the GAMBAS middleware, it is clearly unreasonable to assume that all data providers will immediately switch their data format. Consequently, the GAMBAS middleware provides ways to integrate legacy data that does not follow the linked open data principles. To do this, the GAMBAS middleware pursues a dual strategy. For frequently used personal data coming from different existing services such as Google calendar or Facebook, the GAMBAS middleware provides fully functional wrappers that allow the use of the stored information in order to compute privacy policies or to use them as sensor inputs. For public data coming from existing services such as the route information and time tables of public buses, the GAMBAS middleware provides support by simplifying the development of legacy data wrappers. Together, this allows the immediate use of frequently used data and it fosters extensibility with respect to more specialized existing services.

2.1.3.3 Data Dynamics

Finally, the last dimension categorizes the data on the basis of its dynamics. Intuitively, the dynamics of the underlying data can have a significant impact on the way it needs to be handled by the architecture. Clearly, there is a broad spectrum of possible dynamics and even data such as street names, which can be considered to be static, is subject to change. However, at both ends of the spectrum, we can identify the following categories:

- **Static Data:** Static data is data that never changes or changes rather infrequently. Examples for static data are geographic information such as a map of a city or the route information of a public bus. Clearly, both examples can change over time. However, considering their update rate of months or years, it is usually possible to query the information once and then cache the results of the query for a significant amount of time. Aside from caching intermediate results, it is also possible to replicate the complete set of static data that is offered by one service at another service in order to trade-off storage for network bandwidth and latency. Given this optimization potential, the handling of static data is often less demanding than the processing of dynamic data.

- **Dynamic Data:** Dynamic data is data that changes frequently. Examples for dynamic data are the location of a particular user or a bus in the city. Although there might be periods in which updates are less frequent, like at night when the user is sleeping or the bus is parked at the depot, in many cases, it is not possible to apply similar optimizations as with static data. For example, the application of replication will require frequent synchronization and the introduction of caches for intermediate results may lead to significant imprecisions. Consequently, in many cases, dynamic data requires the execution of continuous queries, which are more resource-intensive to evaluate.

2.2 Dynamic Perspective

Given the introduction of the entities, building blocks and data types in the static perspective of the architecture, the dynamic perspective describes how they interact in order to achieve the different goals. Due to the technical objectives of the GAMBAS middleware, the dynamic perspective focuses on three main parts, namely the acquisition view, the processing view and the inference view. The acquisition view describes how different types of data are collected. The processing view describes how different types of data can be queried. The inference view describes how different data inferences can be drawn using the architecture.

2.2.1 Acquisition View

From the point of view of data acquisition, the GAMBAS middleware supports two different scenarios. The first scenario is targeting the personal acquisition of data that is used to capture the user's behavior on behalf of the user. The second scenario is targeting the collaborative acquisition of data from a large number of users that is used to improve or provide a particular service upon request of a service provider.

For the first scenario, the identity of the user is important to ensure that the resulting profile can be associated with the right user. Consequently, the acquired data may be highly sensitive from a privacy perspective. For the second scenario, the identity of the user is often not important but the service provider is rather interested in an aggregated view of the data. Consequently, by ensuring that the acquired data cannot be associated directly with a particular user, the resulting privacy issues of data collection can be minimized.

Independent of the type of acquisition, we assume that the user must be able to give an explicit consent to the data acquisition at least once in order to ensure that only the desired data types are acquired. To do this, the user must interact with the privacy framework by means of the intent-aware user interface to set the associated preferences. In the following, we outline how both scenarios are handled from an architectural perspective.

2.2.1.1 Personal Data Acquisition

A primary objective of the GAMBAS middleware is to enable the development of behavior-driven services. Intuitively, the realization of a behavior-driven service requires knowledge about the behavior of the service consumers. A key feature of the GAMBAS middleware is to provide support for the gathering of such knowledge automatically in the background.

In contrast to other approaches, the middleware focuses on the use of personal mobile Internet-connected objects such as tablets or smartphones as primary platforms for data acquisition. The reasons for this are manifold. First and foremost, many Internet-connected objects are self-contained and do not require additional infrastructure support. Secondly, the objects are often not utilized to their fullest capacity, leaving enough resources to perform context recognition. Thirdly, many Internet-connected objects have access to both physical and virtual data sources, which allows multi-modal context recognition with high precision. Lastly, the object's context is usually tightly correlated to the user's context and the recognition alone (i.e. without sharing) does not invade privacy.

While the former points are primarily underlining the technical suitability of personal mobile Internet-connected objects as acquisition platforms, the last point highlights a key feature of the approach taken by GAMBAS that is the explicit decision to focus on privacy. Given the possibly privacy-sensitive nature of a behavior profile, the data contained in it must be considered private unless a user actively shares it, e.g., in order to enable service adaptation. Consequently, the data should not be accessible directly by other parties

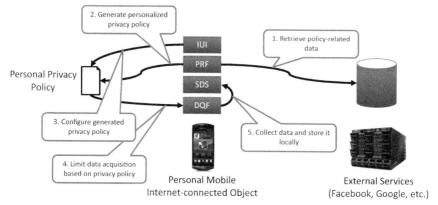

Figure 2.3 Personal Data Acquisition.

rather than the user. In order to achieve this, all personal data collected by the system is stored locally on the devices of the respective users.

The resulting component interaction for personal data acquisition is depicted in Figure 2.3. To reduce the configuration effort for the user, the privacy framework retrieves policy-related data from third-party services such as Facebook or Google, for example. Using this data, it computes an initial privacy policy. This policy can then be refined through the intent-aware user interface in order to enable manual control over all aspects of data acquisition and sharing. The resulting personal privacy policy is then used by the data acquisition framework, which limits the acquisition to those data types that are allowed by the user. In order to limit the access, the acquired data is stored locally in the semantic data storage of the mobile Internet-connected object.

2.2.1.2 Collaborative Data Acquisition

In addition to personal data acquisition, the GAMBAS middleware also supports the collaborative collection of data, for example, to enable the optimization of services based on aggregated usage information. Intuitively, this requires an alternative to the previously described personal data acquisition, since the local storage of data is not suitable for aggregating remote data. To support this, the GAMBAS architecture introduces the ability to remotely store information. Intuitively, this remote storage raises additional privacy concerns since a service provider might be able to associate the reported data with a particular user.

To mitigate this, the GAMBAS middleware enables fine-grained control over the collection process using the same procedure that has been introduced for personal data acquisition. This enables a user to control the data that will be acquired on behalf of a service provider. In addition, the architecture also enables modifications to the data that is reported to a service. As a simple example, the middleware could refrain from sending unique identifiers or could replace them with (randomly) generated pseudonyms that change over time. In more complicated scenarios, the middleware might also apply obfuscation to reduce the data quality or it might refrain from reporting certain pieces of information at all. For this, the data acquisition framework provides a user with control over the data that is reported. This control can then be exercised to limit the sharing of data in such a way that it does not conflict with the users privacy requirements.

The resulting component interaction for collaborative data acquisition is depicted in Figure 2.4. Like in the personal data acquisition case, the privacy framework retrieves policy-related data from third-party services, which is used to compute an initial privacy policy. This policy can then be refined through the intent-aware user interface in order to enable manual control. The data acquisition framework uses the resulting privacy policy to limit the data acquisition to those data types that are allowed by the user and to modify the data accordingly before transmission. As a last step, the data acquired by the adaptive data acquisition framework is then sent to a remote device where it is stored or further processed.

2.2.2 Processing View

To describe the processing of queries, it is necessary to consider the different classes of data depending on the possible level of access. Intuitively, since

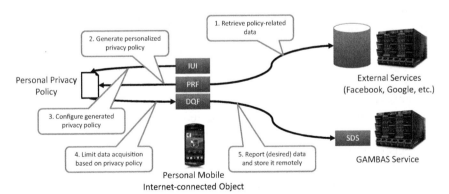

Figure 2.4 Collaborative Data Acquisition.

private data is only available to the device that collected it, distributed processing is not possible with private data. Instead, only local queries can be executed on it. However, as described previously, a user may share (parts of) his/her private data with other users or their devices. In order to ensure that shared data can only be accessed by legitimate entities, an associated access control mechanism is required. Furthermore, it is necessary to encrypt the underlying communication in order to avoid the overhearing of data over insecure network connections. For public data, access control and encryption are not necessary since the data is freely shared with everyone. Due to these differences, the GAMBAS architecture supports two possible data discovery and access mechanisms that are used depending on the level of access granted to the data. In the following, we describe both of them individually. Intuitively, it is possible to create queries that involve public as well as shared data by combining both approaches. Similarly, local queries follow the same idea but since they are targeting only data that is available locally, the associated discovery procedures are omitted.

2.2.2.1 Processing of Public Data

Overall, the processing of public data relies on the following generic three-step procedure that is frequently used in service-oriented architectures:

- **Export (Announcement, Publication):** In the first step, the availability of the data is indicated to other devices by means of exporting metadata (which describes the available data) to the data discovery registry. Depending on the architecture, this step is often referred to as export, announcement or publication. If the underlying data changes in such a way that the metadata is no longer valid, the changes must be reflected by an update to the exported metadata to avoid stale references.

- **Search (Lookup, Binding):** In the second step, which takes place before query execution, the data discovery registry is used by the query processor to find the relevant data sources. To do this, the query processor executes a query on the metadata that is stored in the data discovery registry. The query that must be executed on the metadata typically depends on the query that has been posted to the query processor from an application. Based on the result of the query against the data discovery registry, the query processor continues with the execution of the actual query against some of the retrieved data sources.

- **Execution (Usage, Invocation):** In the third step, the actual query is executed against the data sources. Depending on the capabilities of the query processor, the query execution might be decomposed in further phases such as query planning and query execution. In the query planning phase, the query processor will typically select one of multiple possible query execution strategies in order to optimize certain goals such as decreasing the network load or decreasing the resource usage on certain types of devices.

Figure 2.5 shows an example for the execution of a query on two public data sources. Intuitively, steps one, two and three are decoupled in time, i.e. they must happen sequentially but the time period between them may vary.

As a first step, the public data sources announce their data by exporting associated metadata to the data discovery registry. Typically, this is done once the device starts up its semantic data storage and the announcement might be updated in cases where the data storage holds dynamic data that is reflected in the metadata. Intuitively, however, the update frequency of the metadata should be lower than the update frequency of the actual data in order to avoid scalability issues with the data discovery registry. Once a query is issued, for example, through the intent-aware user interface, the query processor receives it and interprets it. Based on its contents, it will then create and execute queries on the data discovery registry, which results in a set of possible data sources. Based on the strategy taken by the query engine, an appropriate query plan is generated and executed. For the execution, the query processor executes sub-queries against the necessary set of semantic data storages – via their local query processors – and returns the result to the intent-aware user interface.

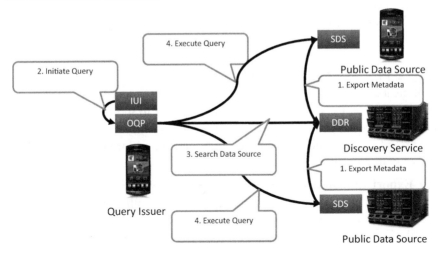

Figure 2.5   One-time Processing of Public Data.

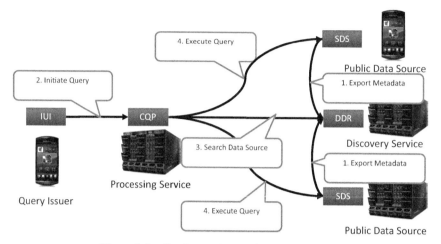

Figure 2.6   Continuous Processing of Public Data.

While the approach described above is sufficient to enable the processing of one-time queries, the execution of continuous queries over dynamic data raises additional issues. Due to the associated resource requirements, the GAMBAS middleware does not encompass a continuous query processor for all types of devices. Instead, the continuous query processor is only suitable for Traditional Computer Systems (TCS) and Backend Computer Systems (BCS). Consequently, it is necessary to handle continuous queries on other systems by means of a third-party system. For public data, this third-party system can be introduced easily. As depicted in Figure 2.6, the system simply acts as a proxy for query processing and there is no need to change the remaining interaction.

2.2.2.2 Processing of Shared Data

As indicated before, the processing of shared data cannot be handled in the same manner as the processing of public data due to the additional require-ments on access control and encryption. Consequently, we need to modify and extend the previous interaction by introducing additional steps that take care of both. For this, the architecture foresees the following general process:

- **Export (Announcement, Publication):** As with public data, the first step is to announce the availability of data to other devices by means of exporting metadata. However, in contrast to public data, only the device identity will be exported in order to avoid privacy issues resulting from the export of private metadata. In cases where no privacy issues result, other metadata could be exported as well in order to improve the performance of the query processor. Alternatively, it is possible to encrypt the metadata as described in Chapter 4.

- **Search (Lookup, Binding):** In the second step, which takes place before query execution, the data discovery registry is used by the query processor to find the relevant data sources by means of querying their identities.
- **Preparation:** The third step takes care of the creation of a view of the remote data that shall be shared with the device that executes a query. The view creation itself consists of a number of sub-steps. First, the identity and data requirements are forwarded to the privacy framework of the device issuing the query. Second, the privacy framework contacts the privacy frameworks on the devices hosting the shared data. For this purpose, the privacy framework performs a mutual authentication. Furthermore, the privacy framework executing on the devices hosting the shared data performs access control, which will eventually result in the creation of a view that represents the data that shall be visible to the requester. Thereby, it is noteworthy to mention that this view may modify the original data based on the level of access. For example, the device hosting the data might decide to generalize parts of the data or to make parts of the data inaccessible. Once the view is created, a secure token is generated, which can then be used to access the view. This token is returned back to the query processor.
- **Execution (Usage, Invocation):** In the last step, the actual query is executed against the view provided by the devices hosting the shared data. In order to access the view, the query processor provides the token to the shared data source and it uses an encrypted channel to transmit both the query and the result.

Figure 2.7 depicts this process with one device that is issuing a one-time query on two sources providing shared data. As described previously, the time between export and access of the device's identify information in the data discovery registry may be high since the device providing shared data will usually export its identity as well as other public and optionally encrypted metadata that does not raise privacy issues upon startup.

Following the general process described above, the devices hosting the shared data export their identity and non-privacy-critical metadata in a similar fashion, as public data sources will share their metadata. As depicted in Figure 2.7, the processing is then initiated by means of a query issued by

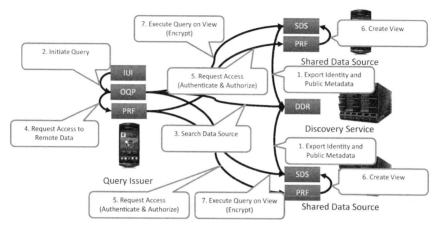

Figure 2.7   One-time Processing of Shared Data.

the intent-aware user interface. In order to access shared data, the query will typically specify the identity of one or more data sources whose connectivity information must be retrieved by searching through the data discovery registry in step two. Once the information is retrieved, the query processor will request the creation of the view through the local privacy framework. To prepare the view, the privacy framework on the query issuer contacts the devices hosting the shared data. Thereby, the privacy framework components on the devices will jointly perform request authentication and authorization. If this is successfully completed, the devices hosting the shared data will create the view on the data that shall be exposed to the query issuer. Thereby, they may perform arbitrary operations on the data such as generalizing information or removing information from the view on a per-request basis. Once the view is prepared, the privacy framework on the query issuer device will receive an access token enabling it to access the newly created view. This token is then passed back to the query processor, which will then issue the respective sub-queries to each of the data sources (again via the local query processors). Thereby, the whole transaction is encrypted and authenticated using the token. Once the sub-queries have been executed, the views on the devices hosting the shared data will be disposed and the result will be returned to the intent-aware user interface.

To enable stream processing on Constrained Computer Systems (CCS), the architecture mimics the proxy-based approach taken for public data where a remote processing service provides the hardware and software resources to perform queries. As shown in Figure 2.8, the primary difference between

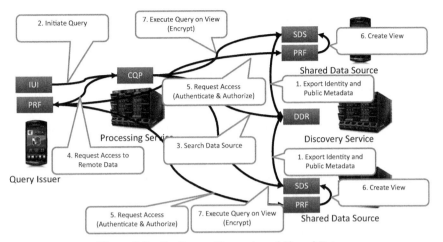

Figure 2.8 Continuous Processing of Shared Data.

continuous and one-time processing originates from the fact that a third party will be involved in data access. Consequently, this third party will have to be trusted by the shared data sources or they have to deny the request. In order to make the associated decision, the access request will not only have to authenticate and authorize the query issuer but it also has to decide upon the trust of the processing service. To do this, the request must identify the processing service that will be used during the processing. Furthermore, the remote communication between the actual query issuer and the continuous query processor must be secured accordingly. However, as described in Chapter 5, this can be done using regular cryptographic methods.

2.2.3 Inference View

Another key usage scenario of the GAMBAS middleware is the gathering of data from multiple sources in order to derive additional information. Based on the scenarios described in Section 1.3 and considering the acquisition and processing views described in the previous sections, we can identify two main classes of inferences, namely local and distributed inferences.

Local inferences are primarily based on information that is available to a single device. Thereby, the term local does not exclude the use of static public or shared data. A device can opt for locally storing a view of the static remote data to simplify the processing.

For dynamic data, however, such an approach is usually ill-suited – at least for devices that cannot host a continuous query processor – since the remote data would have to be refreshed continuously. Consequently, from an architectural perspective, the GAMBAS middleware realizes (simpler forms of) distributed inferences by means of executing continuous queries. Intuitively, the degree to which such inferences are possible depends on the capabilities of the query language. In the following, we briefly outline the approach taken to support local and distributed inferences.

2.2.3.1 Local Inferences

As indicated previously, local inferences encompass inferences on local data that is held by a device as well as public or shared static data that is available remotely and can be accessed by the device. Given the local availability of data, it is then possible to derive additional data using custom software that is executed by the device. Based on the application scenarios described in Section 1.3, the GAMBAS middleware supports two types of such local inferences which we outline in the following.

- **Personal inferences:** Personal inferences are inferences over the behavior data collected using personal data acquisition. They may entail the derivation of aggregated information from multiple sensors or the prediction of future behavior based on traces of the past behavior. In order to enable such an aggregation or prediction, the device itself may require additional static information. For example, in order to determine the typical bus stops that are used by a user, the personal mobile object of the user may have to retrieve the GPS coordinates of the bus stops. Similarly, in order to predict future trip destinations, it may be necessary to retrieve the street address of previously visited locations. From an architectural perspective, the GAMBAS middleware provides support for such personal inferences mainly by means of the adaptive data acquisition framework, which provides extensible support for data aggregation, behavior tracking and prediction. In addition, it is also possible to support this type of inferences at the user interface level in cases where the required inferences are highly application-specific.

- **Service inferences:** In addition to personal inferences which affect the data stored on the personal mobile objects, there is also a need to support service-specific inferences, which may be a result of the aggregation of data acquired collaboratively. As a simple example, it may be necessary to aggregate collaborative sensor readings in order to derive an environmental map. Similarly, it might be necessary to assign the occupancy information collected by different users to different buses on a route in order to predict their remaining capacity. Intuitively, such aggregations should be handled by the service that collects the data since they may require the combination of data reported by multiple sources. However, since such aggregations are often highly application-specific, the GAMBAS middleware primarily supports them by providing data access by means of a generic query processor, which is detailed in Chapter 4. For aggregations that exceed the capabilities of this query processor, it is necessary to implement custom application logic.

2.2.3.2 Distributed Inferences

In contrast to local inferences which are limited to local data and static remote data, distributed inferences combine the dynamic data of multiple remote data sources. Considering the application scenarios targeted by the GAMBAS middleware, such inferences are necessary, for example, in order to detect the collocation of friends in the same public bus. Since we can assume that many users are not willing to publicly share their current location, such information is typically only shared among a specific and user-dependent set of people. To protect the privacy of the users, the GAMBAS middleware only stores this information on the personal mobile objects of the associated user. Thus, given that the user is willing to share this information, it must be retrieved from there.

As indicated previously, however, due to the resource constraints of many personal mobile devices, the GAMBAS middleware does not encompass a continuous query processor for these types of devices. Thus, there are only two ways of supporting distributed inferences. The first and simplest way is to perform the inference by means of a continuous query that is executed on some third-party system that is jointly trusted by the group of users. The second way to realize such inferences is to implement a custom service to perform the inference.

Both approaches have different benefits and limitations. The former approach does not require any custom implementation and thus, is easier to realize. However, given that the query language applied in GAMBAS may not support arbitrary application-specific functions, it is limited to the set of operators supported by the language. The latter approach does not suffer from this problem since arbitrary code can be used during the service implementation. However, in contrast to the use of an existing middleware component, it is much more complicated as it requires the development of custom software. To realize the applications described in Chapter 6, it was sufficient to use a continuous query processor in order to perform the desired distributed inference; however, in a broader context, it may be necessary to provide additional custom services.

2.3 Interface Perspective

Given the static and dynamic perspective of the high-level architecture described in the previous sections, it is possible to identify the interfaces between the core building blocks that constitute the GAMBAS middleware. To do this, we take the different building blocks detailed in the component view of the static perspective, namely the data acquisition framework (DQF), the semantic data storage (SDS), the one-time and the continuous query processor (xQP), the data discovery registry (DDR), the privacy preservation framework (PRF) and the intent-aware user interface (IUI). Using these building blocks, we step through all interactions described in the dynamic perspective, namely the acquisition view, the processing view and the inference view. Consequently, we get the following interactions:

- **DQF–SDS:** In order to store contextual information for personal as well as collaborative data acquisition, the data acquisition framework needs to interact with the semantic data storage.
- **xQP–SDS:** To execute one-time as well as continuous queries over public or shared data, the one-time and continuous query processor needs to interact with the semantic data storages hosting the data.
- **xQP–xQP:** To execute distributed queries, the query processor needs to communicate with remote instances of itself to place subqueries there and receive results back.
- **xQP–DDR:** In order to determine the appropriate data source before executing a distributed query, the one-time as well as the continuous query processor needs to interact with the data discovery registry.
- **xQP–PRF:** When executing a distributed query over shared data, the one-time as well as the continuous query processor needs to interact with the privacy framework in order to gain access to the shared data.
- **PRF–DQF:** To protect the user from unwanted data collection, the access to the data acquisition framework is guarded by the privacy preservation framework. Thus, performing data acquisition requires interaction between these components.
- **PRF–PRF:** In order to gain access to shared data, the privacy framework must be able to interact with other instances of the framework remotely

to negotiate the appropriate access levels and to prepare the necessary views for querying.

- **IUI–xQP:** To retrieve data from the local storage or from remote services, the intent-aware user interface must execute distributed one-time or continuous queries over the associated data storages using the one-time and the continuous query processor.
- **IUI–PRF:** Although, the privacy preservation framework is supposed to enable the automated generation of a privacy policy, some users may want to control their sharing more tightly. In order to enable this, the intent-aware user interface enables manual configuration of the privacy framework.

In the following subsections, we describe the required functionality and the resulting interfaces in more detail. Thereby, we try to refrain from describing low-level implementation details. Instead, we rather focus on an architectural perspective by describing the types and flows of data that is interchanged through the interfaces. However, where appropriate, we also give some initial ideas on how this flow can be realized.

2.3.1 Storage Interfaces

DQF–SDS Interface: As detailed in the acquisition view, there are two basic types of data acquisition foreseen in the GAMBAS architecture, namely personal data acquisition (on behalf of the user) and collaborative data acquisition (voluntary, on behalf of a service). For both types of data acquisition, it is necessary to store the sensed data – either on the mobile device or on a remote device hosting the service. To perform the data acquisition, the GAMBAS architecture introduces the data acquisition framework (DQF) which is responsible for recognizing different types of context in an energy-efficient manner. To store data following the open linked data principles that are forming the core of the interoperability mechanisms provided by the GAMBAS, the architecture foresees a semantic data storage (SDS). In order to enable the persistent storage and the later retrieval of recognized context, it is necessary to introduce an interface between these two components.

Based on this rationale, it is necessary to support the insertion of data into the semantic data storage. In addition, in order to support the storage of transient states, it is also necessary to support the deletion of data from the storage. This allows, for example, the removal of stale entries. As described in detail in Chapter 4, the storages are using RDF as their internal data format.

Thus, the insertion and deletion functions or the storage use this format as well. As a result, the interface consists of the following two functions that can be called either locally (to support personal data acquisition) or remotely (to support collaborative data acquisition):

- **Insert (RDF Triple) Success :: Local & Remote:** Enables the insertion of an RDF triple into the semantic data storage by placing either a local or remote call and indicates whether the insertion has been completed successfully.
- **Delete (RDF Triple) Success :: Local & Remote:** Enables the deletion of an RDF triple from the semantic data storage by placing either a local or a remote call and indicates whether the deletion has been completed successfully.

As indicated previously, the architecture foresees the usage of the interface through the data acquisition framework by means of a special storage component that enables the application developer to define the data storage that will receive the insertion (or deletion) as well as the graph (i.e. the set of triples) that shall be generated (or updated). Once the data acquisition framework generates a new result and transmits it to the storage component, the storage component will perform a deletion of previously inserted triples (if desired) and execute an insertion of the newly created graph. If the insertion or deletion shall be executed on a remote system (to perform collaborative data acquisition), the same set of procedures shall be executed. To improve the overall performance of the interface, in particular, when executing insertions and deletions remotely, it is beneficial to support batch insertions and deletions. This can significantly reduce the latency of updates, especially when multiple triples have to be removed and inserted into a remote storage over a low-bandwidth connection (such as a GPRS link, for example).

2.3.2 Query Interfaces

xQP–SDS Interface: As explained in the components view, there are two types of query processors. One-time query processors (OQP) aim at executing one-time queries, i.e. queries that are evaluated against the current state of the data. One-time query processors are focused on more static data. For the dynamic data, continuous query processors (CQP) are in place. CQPs can monitor the input data coming from streams, and as soon as a new data item is generated, the CQP will evaluate the query and if new results are produced, they are then forwarded to the query initiator. In both cases, the data is stored in semantic data storages and the goal of the query processors is to make the data from an SDS available to services and applications. Therefore, an interface between xQP and SDS is needed. Query processors should be added to retrieve data from an SDS that matches a query. Additionally, an xQP also servers as an interface to add or delete data in an SDS, for example, the data generated from the user intention interface.

For query optimization purposes, the query processor might make use of what we call "temporary data". Temporary data is used only during the query execution and can be discarded afterwards. As an example, the query processor might decide to temporarily store the public data in a local SDS to avoid remote calls during the processing. To keep the storage costs low, this data would be removed after the query is executed. For this, the query processor need to provide functions to add, retrieve and delete temporary data from an SDS.

Based on this rationale, it is necessary to support retrieving data from a semantic data storage that matches a query, as well as data insertions and deletions. This applies for both persistent data and temporary data. As described in detail in Chapter 4, the data retrieval is done via SPARQL queries. As a result, the interface consists of functions described below. Since the query processing supports the aggregation of data from different sources, the functions can be called either locally or remotely:

- **Insert (RDF Triple) Success :: Local & Remote:** Enables the insertion of an RDF triple into the semantic data storage by placing either a local or remote call and indicates whether the insertion has been completed successfully.
- **Delete (RDF Triple) Success :: Local & Remote:** Enables the deletion of an RDF triple from the semantic data storage by placing either a local or a remote call and indicates whether the deletion has been completed successfully.
- **Insert Temporary (RDF Triple) Success :: Local:** Enables the insertion of an RDF triple into a temporary graph in the local semantic data storage and indicates whether the insertion has been completed successfully.
- **Reset Temporary () Success :: Local:** Deletes all entries in the temporary graph.
- **Retrieve (SPARQL Query) result set :: Local:** Enables to query the local SDS for data items that match a given search pattern. To specify the search pattern, an SPARQL query can be used. Matching data items

are returned as a result set, containing all suitable bindings for each requested variable.

- **Retrieve Temporary (SPARQL Query) result set :: Local:** Enables to query the temporary graph of the local SDS for data items that match a given search pattern. To specify the search pattern, an SPARQL query can be used. Matching data items are returned as a result set, containing all suitable bindings for each requested variable.

The architecture foresees the usage of the interface as in most query processing systems. Since the SDS uses a graph data structure to store the RDF triples, the retrieval function works by finding sub-graphs on the SDS that matches the input query. The query processor takes care of parsing the input query to generate the execution plan. The processor also contains a component that inserts and deletes data from a data storage. To improve the overall performance of the interface, as in the case of the DQF–SDS interface, batch insertions and deletions are a useful optimization, especially on remote calls. Optimizations on the query execution plan, with the use of temporary data are also possible.

xQP–xQP Interface: In the GAMBAS architecture, some queries can only be answered by combining data from multiple sources. One solution would be to gather all data from the relevant sources in a single device and execute the query locally on that device. However, there are many problems with this approach. For starters, it would create a lot of data traffic, since it is not possible to know a priori which data is needed, therefore each source would ship all its data to a single device. Scalability would also be an issue, since the device executing the query would become a bottleneck. Finally, this approach does not preserve privacy and therefore becomes unsuitable for the GAMBAS framework.

Our solution is to equip each device hosting data with a query processor. Each query processor can execute queries locally over the device's data, and it can also aggregate results from multiples sources. For executing a distributed query among the devices, the query initiator first identifies the relevant sources using the DDR. It then breaks the query into subqueries. Each subquery is sent to the device that contains the data for it. The query processor on each device will then execute the subquery locally and only forward the relevant results to the query initiator (as opposed to all data). The query initiator merges the results of all subqueries and creates the final query result.

To execute distributed queries, the query processor needs to communicate with remote instances of itself to place subqueries there and receive results back. This is done by implementing an interface that allows query processors to post queries to remote query processors and retrieve the results, as shown below.

- **Retrieve (SPARQL Query) result set :: Remote:** Enables to query a remote xQP for data items that match a given search pattern. To specify the search pattern, an SPARQL query can be used. Matching data items are returned as a result set, containing all suitable bindings for each requested variable. This is used to place subqueries that are part of a distributed query.

The middleware architecture foresees the usage of the interface during the execution of distributed queries. Since the local execution is done over RDF triples, the interface between query processors is done using a language suitable for RDF, in our case SPARQL.

xQP–DDR Interface: To answer queries which involve remote data, the query processor must be able to discover the data sources that are available on the network. Once a query is issued, the query processor receives and interprets it. This allows the processor to identify which data is needed to answer the query (for example, the location of friends of a user). The data discovery registry contains the meta information about the data sources, not the data itself. This is to preserve the privacy of shared data. By consulting the registry, the query processor can obtain the list of sources that contain the data in question. For instance, the registry can return the list of semantic data storages from the friends' devices.

Based on this rationale, an interface between query processors and registry is needed. The interface must allow the processor obtain a list of remote SDSs (or endpoints) that contains a particular type of data. This interface requires only one functionally, which is given below:

- **Resolve (data source specification) endpoints :: Remote:** Enables the discovery of SDS endpoints that can be contacted for a specific kind of data, e.g. whom to contact to get information about a user's location. To do so, a data source specification is given, e.g. specifying the user for which data is searched (for instance, a friend). This request is sent to the remote discovery server and a set of matching endpoints is returned.

The interface functionality detailed is used during the processing of queries that involve remote data. The query processor identifies which data sources are needed (e.g. the sources containing the location of friends), and request them to the registry. The registry then performs a lookup on the metadata it stores and returns the list of remote storages that matches the request. To improve the performance of the interface, and the performance of the query processing in general, the metadata stored in the registry can be enhanced in order to provide more accurate and results sets. However, storing more metadata might lead to privacy issues, so this needs to be handled carefully.

2.3.3 Privacy Interfaces

xQP–PRF Interface: During the query execution, the query processor identifies the sources needed to answer the query and then sends a request to the registry. The registry resolves the sources and sends the list of endpoints (remote storages) that contain that data in need back to the processor. For shared data however, before the query processor can access the data on the remote source, a privacy control is performed to check if the query initiator has the rights to access the data. A view of the data matching the privacy rules in place is created and shared with the query processor. This is done in the preparation phase explained in Section 2.2.2.2. The query processor forwards the identity and data requirements to the privacy framework, which in turn checks with the privacy framework of the device hosting the shared data. A view of the data is created based on the access control. The view can reflect the original data, or it can modify the original data according to the privacy in place. For example, it can aggregate or hide parts of the original data. Once the view is created, a secure access token is generated and sent to the query processor. If a remote endpoint is trying to access the shared data, the secure access token will allow transferring the shared data securely over the chosen communication channel.

Based on this rationale, an interface between xQP and PRF is needed to check whether the query initiator is allowed to access the data. Additionally, if the xQP is executing a remote query, the communication must be properly secured. For this, the user and data access credentials are sent over a secure proxy that is part of the communication subsystem of the middleware to provide a secure data connection between the two endpoints. As discussed in Chapter 5, the secure proxy manages the secure communication transparently. Thus, the interface does not include a method that enables the exchange of security tokens or start the encryption. In contrast to that, the access to data must be checked through an interface. The interface consists of one function that checks if the query initiator (i.e. the user requesting the data) is allowed to access the data. The data being requested also needs to be specified. The PRF looks at these two input items and decides whether the query is allowed or not. Each request is handled by the privacy framework of each semantic data storage and therefore this function is performed locally. The function supported by this interface is given below:

- **Check (Set of Ontology Classes, Requester) allowance :: Local:** Enables to check with the PRF, if executing a received query is allowed according to the currently active privacy policies. To do so, the query processor hands the PRF (1) a set of classes in the GAMBAS ontology that specify what data types the query will access and (2) the origin of the query, e.g. if it was a local query or a query from a remote user. The PRF returns whether this query is allowed or not.

The architecture foresees the usage of the interface described above in the privacy-preserving query execution mechanism, when shared data is involved. The query processor must first interact with the privacy framework, which is responsible to allow or deny data access and responsible for data encryption/decryption.

PRF–DQF Interface: The Adaptive Data Acquisition Framework (DQF) enables the collection of data using various sensors built into the user's mobile device. The collected data can then be used personally (i.e. by the device, in the case of personal data acquisition) or collaboratively (i.e. by a remote service, in the case of collaborative data acquisition) to optimize services in a behavior-driven manner. Clearly, the data acquired by means of sensors built into the device of a user may raise privacy concerns. Furthermore, the preferences with respect to privacy may vary drastically from user to user. In order to empower users to exercise control over which data can be collected, the access to the data acquisition framework is guarded by the Privacy Preservation Framework (PRF). Thereby, all accesses made to the data acquisition framework are checked against the user's privacy preferences with respect to data collection. This allows the user to limit the data types that can be collected at all. In extreme cases, a user may limit the collection of all data through the GAMBAS middleware. In less extreme cases, the user may limit the collection of a particular type of context information, such as location-related information or audio information.

The PRF–DQF interface enables the data acquisition framework to check whether the user has given consent to the acquisition of a particular type of contextual information. To do this, the DQF performs calls to the PRF in order to verify that the data types that shall be captured are permissible under the user's current preferences. Furthermore, since the user's preferences may change at any point in time, it is necessary that the PRF provides functionality to signal a change to the DQF whenever the user's preferences with respect to a particular data type change. Consequently, the interface must be composed of the following two functions:

- **Check (Datatype) authorize :: Local:** The PRF checks the data type that is about to be captured against the preferences of the user and returns a Boolean to indicate whether the user permits the acquisition of the specified data type. If the access is denied, the acquisition is aborted. If access is granted, the acquisition task can be started.
- **Signal (Datatype) void :: Local:** The PRF signals a change to the preferences with respect to a particular data type such that the DQF can check all currently executed data acquisition tasks against the updated set of preferences. If a data acquisition task is no longer permitted by the user, it must be aborted.

In order to guarantee that all data acquisition tasks continuously conform to the user's preferences, the architecture foresees the continuous and gapless usage of this interface for all calls to the DQF. This means that all tasks that are started within the DQF need to pass through the check method of the PRF with the associated data types. In addition, as long as the DQF is executing any tasks, it needs to react to changes indicated by the signal method. If a signaled change affects a data type that is currently acquired, the check for the associated (set of) task(s) needs to be reevaluated, possibly aborting any conflicting tasks. The check of the DQF against the policy managed by the PRF may entail some slight overhead, which may become significant if data acquisition tasks are started and stopped very frequently. In this case, it makes sense to cache the user's preferences in memory to reduce the associated overhead. However, in most usage scenarios, the overhead can be neglected.

PRF–PRF Interface: The PRF allows the transfer of data between two devices. The data that is transferred should be encrypted. The reason for this is twofold. At first, the data might contain private information that should not be shared with unauthorized users or devices. Additionally, the shared data might be transferred over an insecure communication channel (e.g. the Internet or an insecure WiFi network). To enable encrypted communication, it is necessary for both communication endpoints to use a cryptographic key. Using the efficient concept of symmetric encryption, the key must be identical and exchanged before the secure communication can take place. During the exchange of a cryptographic key, the communication endpoints show that they are eligible to access the data that should be transferred by authorizing themselves. After the authorization process, both endpoints possess a shared cryptographic key that allows them to transfer data securely.

The PRF–PRF interface allows the authorization of communication endpoints. The successful authorization can be performed in two different ways. The first way uses asymmetric cryptography and is based on certificates, similar to the implementation of SSL in the Internet. This allows an ad-hoc identification of devices that belong to a certain domain. If the domain root is trusted, the authorization will be successful. Also, the access rights depend on the trust in this root. For authentication, the device's certificate is transferred together with a challenge that proves that the device is in possession of the certificate's private key. Together this data forms the device's credentials that are checked at the other endpoint. The alternative of using compute intense asymmetric cryptography is symmetric cryptography. Using symmetric cryptography, a key can be attached to a connection between two endpoints. The first half of this shared key allows the identification of the other endpoint. The other half can be either directly used for the secure communication or used to exchange a new session key securely. For efficiency reasons, both of these checks (i.e. for asymmetric and symmetric cryptography) are performed directly during the communication. The local interface is designed as follows:

- **Check (credentials, user pseudonym) authorize :: Local:** The PRF checks the security credentials of a user and returns a Boolean that shows if the user was authorized successfully.

The middleware architecture foresees the usage of the interface described above for every secure transmission of data. The communication endpoint must first authorize each other at the remote privacy preservation framework, before a key for the secure communication is computed. Intuitively, the authorization that is performed by the privacy-preserving framework incurs some overhead during the data transfer. However, without the authorization, the communication partner is unknown to another device and this contradicts the privacy of the transferred data. While the authorization therefore is a crucial mechanism, it is may be possible to use more lightweight security mechanism, resulting in a decrease of the security level, in application scenarios that permit this.

2.3.4 Control Interfaces

IUI–xQP Interface: The Intent-Aware User Interface (IUI) is connected to the GAMBAS middleware through a query processing interface, which provides access to local and remote data sources. Local data pertains, for example, to personal travel information, which may include the user's travel history for making predictions to adapt the IUI to his future travel behavior. Remote data could include transport information hosted by third-party services such as a city's local transport agency (e.g. estimated time of arrivals), time tables and information about the travel habits from the user's friends in the social network as stored on their mobile devices. Since all this data is represented based on linked data principles using RDF triples, it can be queried in a uniform manner by means of a powerful graph-based query language irrespective of what specific kind of data is requested and where this data is located.

When the IUI needs to access data, it uses an interface from xQP to connect to external services and read information objects. In particular, this interface is a facade from xQP that calls methods and translate the received data into an understandable format for IUI. As a result, the interface consists of a single power query processor, which allows us the IUI to specify generic queries over data stored on local and remote SDS:

- **Select (SPARQL Query) result set :: Local & Remote:** Enables the retrieval of bindings for requested variables. To specify the variables as well as conditions that bindings for them must match, an SPARQL select query can be given. A query can be executed on a single data source or multiple ones, allowing to query and integrate information from multiple users at once. Matching data items are returned as a result set, containing all suitable bindings for each specified variable.

Frequent data access may be a critical factor for the IUI, especially when the queries need to be forwarded to remote data storage over cellular network connections. The low bandwidth of these connections and high variance in quality of service may slow down the query process and cause significant delays in information delivery that can negatively affect the user's experience. In order to improve upon this, the design of IUI foresees a caching strategy, where some static data (e.g. routing information, bus coordinates, time tables) is kept on the mobile device so that no repeated updates are required. This is especially useful for transport network data, which is not expected to change very often. For dynamic data (e.g. arrival time or crowd level of a vehicle), possible optimization strategies include primarily pre-fetching, where the data is retrieved before it is requested by the user. As the data is already available on the device prior to the access to the information, the delay experienced by the user can be minimized.

IUI–PRF Interface: The goal of the privacy preservation framework (PRF) is to protect the user's privacy by providing security mechanisms that enable the secure and authentic interaction between different devices. Thereby, access to different types of data is controlled by the privacy preservation framework on behalf of the user. To do this, the privacy framework relies on a policy that defines the user's preferences with respect to the sharing of data with other users. Although the privacy framework attempts to minimize the configuration effort for the user by deriving a suitable policy from the policies that a user is already applying on different social services, there might be cases where the user wants to exercise full control over the sharing of data. To do this, the privacy preservation framework exposes a configuration interface to the intent-aware user interface (IUI) that provides manual control over the sharing.

To exercise manual control over the sharing of information, the privacy preservation framework enables the intent-aware user interface to (re-) configure the privacy policy. Given that the main entities contained in polices are users and permissions on different data types that express that a particular user may access a particular type of data, it makes sense to expose functionality to manipulate these two entities. To enable the development of a visual representation of the user's current privacy policy, the functionality required to manipulate the policy is additionally augmented with functionality to simply retrieve the current policy. In summary, this results in the following six functions that are available only locally and that are only accessible to the intent-aware UI in order to avoid unwanted modifications.

- **ListUsers() usernames :: Local:** This function enables the intent-aware user interface to list the names of users that have been configured on a particular device. The resulting list of user names can be pruned or extended using the following two functions.
- **AddUser(username) void :: Local:** This function enables the intent-aware user interface to add another user to the list of users that have been configured for a device. If the user is already configured, the method simply returns. If the user does not yet exist, it will be added to the list.
- **RemoveUser(username) void :: Local:** This function enables the intent-aware user interface to remove a previously configured user from the list of configured users. If the user is not configured, the

method simply returns. If the user was configured, the user and all its permissions to access data on the device will be removed.

- **ListPermissions(username) datatypes :: Local:** This function enables the intent-aware user interface to view all the permissions that have been configured for a previously configured user. The list will include all data types to which the specified user will have access. If the user has not been configured, the list of data types will be empty.

- **AddPermission(username, datatype) void :: Local:** This function enables the intent-aware user interface to add a permission for a previously configured user such that the specified user will be able to access data of the specified type. If the user is currently not configured or the user already exhibits a permission to access the data type, the method simply returns. Otherwise, the permission will be added for the specified user.

- **RemovePermission(username, datatype) void :: Local:** This function enables the intent-aware user interface to remove a previously added permission on a specified data type for a specified user. If the permission or the user does not exist, this method simply returns. Otherwise, the permission will be removed and the user will no longer be able to access the specified data type.

The primary intended usage of this interface is the manual manipulation of the privacy policy through a graphical user interface on the device of the user. Thereby, it is important to mention that the access to this interface is intended to be restricted to an intent-aware user interface component that ships together with the GAMAS middleware and that it cannot be accessed through other components in order to avoid unwanted manipulations. Consequently, we envision the creation of one or more list views that show which user has access to what type of data as well as controls that enable the injection of changes to these lists.

3

Data Acquisition

This chapter describes the data acquisition framework of the GAMBAS middleware. The description includes discussions on the framework architecture, including the component system for developing context recognition applications and the activation system for enabling automatic, state-based activation of different configurations. The chapter also provides insight into the design rationale for the system. This includes a discussion of the motivation behind the component-based approach for context recognition, the chosen component model, energy-efficient techniques to perform context recognition on resource-constrained mobile devices, etc. Furthermore, rationale behind the state machine abstraction for the activation system and how energy optimization techniques used in the component system are fully utilized by the activation system is given. Before we discuss the framework, however, we first outline related work and clarify the innovations and research gaps closed by the data acquisition framework.

3.1 Focus and Contribution

Data acquisition is an essential part of any context recognition system. For such systems, data acquisition normally involves acquiring raw data from different types of sensors such as accelerometers, microphones, gyroscopes, proximity sensors, Wi-Fi, GPS, etc. The sensors can be embedded into a single device or alternatively, they can be embedded in different devices that are distributed in the environment. The data acquisition system acquires data from these sensors and pre-processes it before forwarding it to more complex recognition logic. Existing data acquisition systems differ depending on the leveraged resources and on the target application requirements. An efficient data acquisition system should be generic enough to be executable in different settings (different hardware and different application requirements) with little or no tuning. In the following, we briefly review the state of the art for data acquisition systems mainly focusing personal mobile devices like smart mobile phones, PDAs, etc. Thereafter, we identify the gaps in the existing solutions and from these gaps, we derive a list of innovations realized by the data acquisition framework of the GAMBAS middleware.

3.1.1 Data Acquisition Frameworks

There exist a number of context data acquisition systems and frameworks for personal mobile devices [CK00]. These frameworks vary in their characteristics depending on their target applications and operating environments. Examples include [HH10], [DHH07], [BM10], [YTN05], [KZX+11], [LYL+10], [GJAS06], [RMM+10] and [CBSG12]. [HH10] describes a data acquisition framework for on-body sensor networks which runs on resource-constrained embedded systems and is used for human activity recognition. [DHH07] describes a context acquisition framework which allows the collection of raw sensor data from different sensing sources. The framework provides programming abstractions for developers to fetch data from different sensor implementation programs without developing the underlying communication mechanisms for the target platforms. [BM10] describes a service-oriented architecture based data acquisition framework. It allows sensor data fusion with local and external sources to build and manage context-aware services for personal mobile devices in a transparent manner. The framework protects the user's private data by using suitable privacy-preserving policies to handle information in P2P networks. [YTN05] describes a context acquisition framework based on a customized sensing platform named Muffin. Muffin supports a variety of sensors to help detect different types of contexts. The Citron framework running on Muffin uses a black box architecture for context processing and provides parallel processing of different sensor data streams (audio, accelerometer, etc.) to identify the user's context. [KZX+11] combines both on-body sensors and mobile phones for joint context recognition. The main contribution of this work is the provisioning of a framework to support the collaboration of TinyOS-based sensor modes and Android-based smart phones. This work also makes use of online training to improve the accuracy of the classifiers and it can automatically turn off redundant sensing sources to save energy. [LYL+10] describes a continuous sensing engine for context recognition applications. It uses the concept of pipes for different sensing sources (microphone, accelerometer, GPS) to balance out the application requirements and the available resources. [GJAS06] outlines a software architecture and a service for on-body sensors as part of the user's attire. The system is realized using MicaZ motes. Challenges addressed in this work include storage of data, uploading of data, synchronization of data, power management of motes, reconstruction of activity logs, user interfaces, etc. The presented architecture is aimed at the future development of smart attire systems. [RMM+10] is a data acquisition framework for detecting user's social and physiological patterns using smart mobile phones. The system can be programmed using a declarative language to describe user behavior models, action base and knowledge base. The system can be adapted at runtime to activate and deactivate sensors. The recognition is based on GMMs (Gaussian mixture models). The system is aimed at helping social scientist to understand the correlation of user emotions with the places, groups and their activities. [CBSG12] is a collaborative context recognition system for smart mobile phones. The system execution is a two-stage process consisting of stages, namely grouping stage and context recognition stage. In the grouping stage, devices are clustered based on their proximity. Once devices are clustered, they scan the environment and send the raw data for subsequent context recognition to a backend server. In the context recognition stage, the system uses coupled hidden Markov models to model activity and location sequences. The system is aimed at advertisement systems where advertisements are shown based on mutual context and interest of user groups.

3.1.2 Rapid Prototyping Tools

There also exist a number of rapid prototyping tools for expeditious development of context recognition applications. Commonly known tools include [SDA99], [BAL08] and [TRL+09]. [SDA99] is targeted at context recognition with pre-deployed sensors and provides a uniform abstraction for applications to access and use context information by hiding the actual context sensing and interpretation from applications. [BAL08] is targeted towards activity recognition for wearable systems. This toolkit provides functionalities to develop distributed context recognition systems as well as reusable components, parametrizable algorithms, filters and classifiers. [TRL+09] is a data gathering and processing open-source platform targeted towards mobile phones with varying hardware capabilities. It consists of a minimal core that can be extended by plug-ins.

3.1.3 Application-Specific Acquisition

The systems mentioned above are generally used for dealing with heterogeneous sensing sources and providing flexibility for application developers to customize applications in a certain way. However, there exist a number of fine-tuned data acquisition and context recognition systems that can only be used in a narrow set of situations. Examples include [BC09], [MLEC07], [LLEC08], [LPL+09] and [EML+07]. These and many alike systems are manually fine-tuned for particular applications and therefore are able to detect only the fixed set of characteristics. As a result, these systems cannot be adapted to dynamic environments which a user might experience in a daily routine. They use built-in sensors in mobile phones to recognize the required context. For instance, [MLEC07] uses microphones and accelerometers to determine user context which is then injected into social networking websites. [LLEC08] uses accelerometers and microphones to detect road conditions. [TRL+09] uses location sensors to identify road traffic congestions. Sound Sense [LPL+09] uses a microphone to classify different types of sounds in the surrounding. [BC09] is a system aimed at video recording of social events in a distributed manner using mobile phones. The phones are grouped based on the social activity in which their users are involved. For detection of a social activity, a phone at the appropriate location is chosen to record events. At the end of the social activity, all recordings from different phones are combined into one video by a backend server to create a video highlighting important events of the social gathering.

Data acquisition and retrieval of contextual information is a resource consuming process which can have a significant effect on overall system performance for resource constrained personal mobile devices. Over the last years, there has been some work towards devising mechanisms for achieving energy-efficient data acquisition and processing. Examples include [KLJ+08], [WLA+09], [RH10] and [RMJ+11]. [KLJ+08] detects changes in the context data at an early stage. For instance, rather than waiting for the results from the classifier, the system detects changes in sample values at the sensor level. Thereafter, only those samples are further processed which can lead in a context change, whereas [WLA+09] uses hierarchical sensor management strategy to detect user states and state transitions and only fires a transition when a particular transition probability is met. As a result, this reduces the unnecessary execution of unwanted sensors. [RMJ+11] is aimed at computing multiple contexts from multiple sensing sources. The authors have proposed a theoretical model that shows the inaccuracy of estimating multiple contexts from multiple sensing sources. The work also presents a heuristic algorithm for searching the set of sensors to recognize the required multiple contexts.

3.1.4 Contribution

Designing a context data acquisition system is usually driven by the target applications and operating environments. Therefore, such systems are optimized with considerations to their requirements. The above-mentioned systems are similarly aimed at optimizing a particular characteristic, which could be the efficient utilization of available resources or the highly accurate recognition of a particular context or the efficient prototyping of context recognition applications. Looking at the description of these systems reveals a need for a generic yet efficient system that in essence should be a complete framework, which, on the one hand, allows efficient usage of available resource and, on the other hand, supports rapid development of specialized recognition applications with high accuracy. The data acquisition framework in GAMBAS middleware bridges these gaps. It aims at providing a complete solution that meets all the aforementioned objectives. Specifically, the data acquisition framework of the GAMBAS middleware adopts a component-based approach allowing multi-modal context data acquisition. The framework provides an extensive component toolkit for rapid development of new context recognition tasks. Using a component-based solution, the data acquisition framework applies resource-efficient techniques (memory, energy, etc.) with no or little impact on the recognition accuracy. Moreover, the data acquisition framework is executable in distributed settings to enhance the quality of desired context and helps in providing relevant services to different groups of users (depending on their location, interests, etc.). Finally, the framework provides a number of basic components that can be used to build applications. These components cover activity and intent recognition as well as sound and speech recogntion.

The activity recognition components in the data acquisition framework focus on computing various user activities or user contexts. Due to the application scenarios targeted by the GAMBAS middleware, the primary focus lies on location-based activities, e.g. shopping in a supermarket, waiting for the bus at the stop, traveling in a bus (standing or sitting), sightseeing in a new city, etc. The data acquisition components rely on a variety of means (motion sensor, Wi-Fi, GPS, on-line calendars) to recognize these and similar activities. Similarly, the data acquisition framework encompasses necessary components to estimate the user's intents. By this, we mean the user's likely location or activity in the future, e.g. knowing that a user is traveling on a bus to his destination, it might be useful or interesting to notify him about the possibility of meeting a friend. If he is willing to change his route, then prompt him of new shopping facilities near the destination. The intent recognition components support computing such user intents based on user's activity patterns or interests.

The sound and speech recognition components focus on interpreting acoustic signals in the environment of the user. A primary focus lies on the recognition of environmental sounds, like engine sounds, traffic noise, talking people, etc. to determine the means of transportation. The goal is to identify delays in public transportation to adjust the predictions of personal intentions. The so acquired data can be distributed in accordance with the privacy setting to optimize travel plans of other users who rely on the same means of transportation. The components use historic data and compare it to live data to identify differences in schedule or behavior patterns. The speech recognition components are designed to allow the integration into other applications on the device. This allows developers to create new applications that offer voice control via speech recognition.

3.2 Data Acquisition Framework

The data acquisition framework (DQF) is one of the fundamental building blocks of the GAMBAS middleware. Conceptually, the DQF is responsible for context recognition on personal mobile devices including smart phones, PDAs and laptops. The DQF supports various platforms including Android, Windows and Linux. It is realized as a multi-stage system. At lower stages, it allows developing reusable components and component compositions for context recognition applications. At higher stages, it enables application developers to automatically activate compositions when needed. To do this, the DQF is split into two parts as shown in Figure 3.1, a component system and an activation system.

The component system uses a component abstraction to enable the composition of different context recognition stacks that are executed continuously. A context recognition stack or simply a configuration refers to a set of sampling, preprocessing and classification components wired together to detect a specific context. Examples of such contexts include the physical activity of a person, the location of a person, etc. The configurations can be used to detect context for a multitude of purposes and have applications in

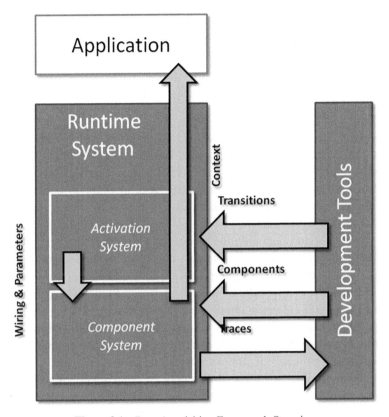

Figure 3.1 Data Acquisition Framework Overview.

areas of smart home environments, assisted living for elderly, proactive route planning, shopping, etc.

The activation system uses a state machine abstraction to determine the point in time when a certain configuration or a set of configurations should be enabled. The activation system enables the required configurations in an automatic manner based on the conditions associated with the state transitions. An example of a simple (coarsely granular) state machine associated with an employee could consist of two states, "Working" and "Relaxing". State "Working" may consist of configurations "Meeting", "Cafeteria", etc. and state "Relaxing" may consist of configurations "Living Room" and "Gardening". Based on the transition values, the activation system will disable the configurations associated with one and enable the ones associated with the other. In addition, the state machines can also have more fine granular states representing stages specific to a single task, e.g. a state can represent the sampling of an accelerometer with lower or higher rate. In such a case, a state change may occur when the device screen turns on, for instance. In the following, we describe both systems in detail.

3.2.1 Component System

At the lower level of the data acquisition framework, context and activity recognition is done using a component-based approach which promotes reusability and rapid prototyping. In addition, this approach also enables the automated analysis of application structures in order to optimize their execution with respect to energy efficiency.

From the perspective of the component system, each application consists of two parts: the part containing the recognition logic and the part containing the remaining application logic. The part that contains the recognition logic usually consists of sampling, preprocessing and classification components that are connected in a specific manner as shown in Figure 3.2. The part that contains the remaining application logic can be structured arbitrarily. Upon start up, a context recognition application passes the required configuration to the component system, which then instantiates the specified components and executes them. Upon closing, the configuration is removed by the component system which eventually releases the components that are no longer required. The component system is implemented in Java and supports various platforms

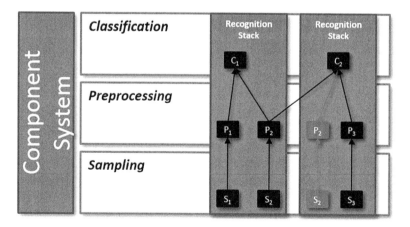

Figure 3.2   Component System Overview.

including J2SE environments and Android. Using an Eclipse-based graphical editor, application developers can visually create configurations by selecting and parameterizing components and by wiring them as needed. In the following, we first provide more details on the underlying component model, before we discuss the runtime and development support.

3.2.1.1 Component Model

To structure the recognition logic, the component system realizes a light-weight component model which introduces three abstractions. First, *components* represent different operations at a developer-defined level of granularity. Second, *connectors* are used to represent both the data and the control flow between individual components. Third, *configurations* are used to define a particular composition of components that recognizes one or more context characteristics.

3.2.1.1.1 *Components*

Components are atomic, reusable building blocks that constitute the recognition logic. Similar to other systems such as J2EE or OSGi, components can be defined at arbitrary levels of granularity. Yet, in contrast, they can be instantiated multiple times and they are parameterizable to support different application requirements. Due to the support for parametrization, the component model is more flexible than other models. In addition to parameters, all components exhibit a simple life cycle that consists of a started and a stopped state. To interact with other components, a component may declare a set of typed input and output ports that can be connected to other components using connectors.

As depicted in Figure 3.3, the recognition logic of a speech detection application may, for example, consist of a number of components which can be divided into three levels. At the lowest level, the sampling components are used to gather raw data from an audio sensor. On top of the sampling components, a set of preprocessing components take care of various transformations, noise removal and feature extraction. Finally, the extracted features are fed into (a hierarchy of) classifier components that detect the desired characteristics. Depending on the purpose and extent of the application logic, it is usually possible to further subdivide the layers into smaller operators. Although the component system does not enforce a particular granularity, such operators should usually be implemented as individual components to maximize the potential for reuse.

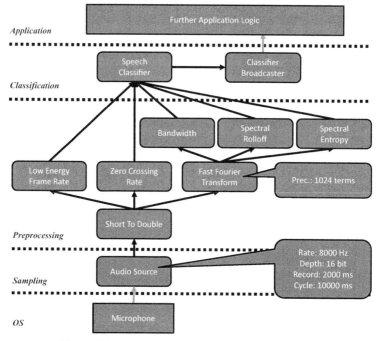

Figure 3.3 Speech Detection Configuration Example.

3.2.1.1.2 *Parameters*

Parameterizations increase the reusability of a component implementation across different applications. The component system allows components to support a developer-defined set of parameters. Components expose these parameters to adapt their internal behavior. As shown in Figure 3.3, at the sampling layer, these parameters might be used to express different sampling rates, sampling depths, frame sizes and duty cycles. At the preprocessing layer, they might be used to configure different filters or the precision of a transformation. In the component system, the parameters are not exposed to other components. Instead, they can be accessed and manipulated by the components.

3.2.1.1.3 *Ports*

In order to support application-independent composition, each component may declare a number of strongly typed input and output ports. Input ports are used to access results from other components. Output ports are used to transfer computed results to another component. Thus, ports enable components to interact with each other in a controlled manner. The developer can add multiple input and output ports of different types. The component system takes care of the necessary memory allocation and de-allocation and performs efficient buffer management for each of the ports in transparent manner.

3.2.1.1.4 *Connectors*

In order to be reusable, components are isolated from each other by means of ports. However, the recognition of a context feature often requires the combination of multiple components in a specific way. Connectors express such combinations by determining how the typed input and output ports of different components are connected with each other. In order to minimize the overhead of the component abstraction, connectors are implemented using an observer pattern [GHJV95] in which the output ports are acting as subjects, whereas the input ports are acting as observers. This enables modeling of 1:n relationships between the components, which is required to avoid duplicate computations. To avoid strong coupling between components, input ports do not register themselves at the output ports, but the component system takes care of managing all required connections. An example of connectors can be seen in Figure 3.3, where the output port of the fast Fourier transform component is connected to the input ports of the bandwidth, the spectral roll off and the spectral entropy component.

3.2.1.1.5 *Configurations*

To recognize a particular piece of context, a context recognition application must explicitly list all required components together with their connectors in a so-called configuration. While this approach slightly increases the development effort, it also increases the potential reuse of components that can be applied on data coming from different sources. As an example of such component, consider a Fast Fourier Transform (FFT) that converts a signal from its time domain into the frequency domain. Clearly, such a component can be applied to various types of signals such as acceleration measurements or audio signals. Thus, by explicitly modeling the wiring of components as part of a configuration, it is possible to reuse this component in different application contexts. In addition to listing components together with their connectors, the support for parameterizable components also requires the developer to explicitly specify a complete set of parameter values that shall be used by each component. As a result, every configuration consists of a parameterization as well as associated connectors. An example of a speech detection configuration is shown in Figure 3.3.

3.2.1.2 Runtime System

The main task of the runtime system of the component system is to support the execution of configurations defined by different context recognition applications in an energy-efficient manner. This includes loading the configurations specified by the context recognition applications, instantiating the components with right parameterizations and connecting them in the manner specified by the application. In addition to that, the runtime system applies energy optimization techniques if more than one application is executed simultaneously. When the applications do not require the context information anymore, the runtime system stops executing the associated configurations. A detailed description of the component system structure and execution of applications is given in the following sections.

3.2.1.2.1 *System Structure*

As shown in Figure 3.4, the main elements of the runtime system of the component system are the configuration store, the configuration folding algorithm [IHW+12] and the applications. The configuration store is used to cache the configurations associated with applications that are active. It is also used to store their folded configuration. The configuration folding algorithm provides energy-efficient execution of context recognition applications, provided that more than one application is executed simultaneously. The entity responsible for managing the runtime system is called the component manager.

3.2.1.2.2 *Configuration Execution*

The component manager controls the execution of the componentized recognition logic of all running applications. To manipulate the components executed at any point in time, the component manager provides an API that enables developers to add and remove configurations at runtime. When a new configuration is added, the component manager first stores the configuration internally. Then, it initiates a reconfiguration of the running recognition logic that reflects the modified set of required configurations. To reduce the energy requirements, the component manager does not directly start the components contained in the configuration. Instead, it uses the set of active configurations as an input for our configuration folding algorithm.

The goal of the configuration folding algorithm is to remove redundant components that are present in different applications and perform the same sampling or compute redundant results. Using the set of configurations, the configuration folding algorithm computes a single, folded configuration that produces all results required by all running applications without duplicate sampling or computation. Once the configuration has been folded, the component manager forwards it to the delta configuration activator. By comparing the running and the folded configuration, the activator determines and executes the set of life cycle and connection management operations (starting, stopping and rewiring of components) that must be applied to the running configuration in order to transform it into the folded target configuration. When executing the different operations, the delta activator takes care of ensuring that their ordering adheres to the guarantees provided by the component life cycle. To do this, it stops existing components before they are manipulated. This procedure is illustrated in Figure 3.4.

3.2.1.2.3 *Platform Support*
The core abstractions of the component systems as well as the component manager are implemented in Java 1.5. In order to support multiple platforms, different wrappers have been implemented that simplify the usage of the component system on platforms including Windows, Linux and Android.

3.2.1.3 Tool Support
The component system encompasses offline tools to support rapid prototyping. These tools include a visual editor which is used for creating and

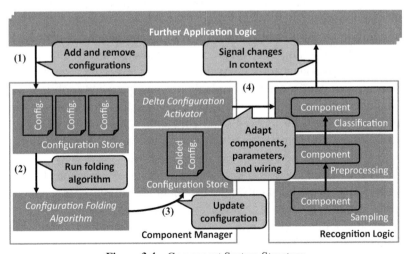

Figure 3.4 Component System Structure.

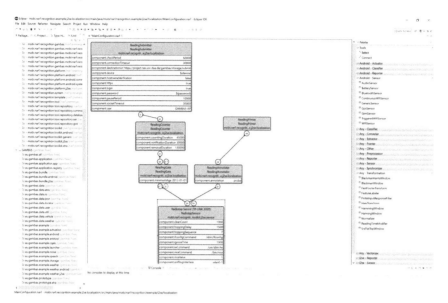

Figure 3.5   Component System Tool Support.

updating configurations for the context recognition applications. The visual editor provides a user-friendly interface, which allows developers to drag, drop, parameterize and wire existing components to create new configurations or update existing ones. The visual editor is implemented as a plug-in for the Eclipse IDE (Version 3.7 and above). A screenshot of the visual editor is shown in Figure 3.5.

In addition to the visual editor, the component system also provides a large set of generic sampling, preprocessing and classification components as part of the component toolkit. At the sampling level, the toolkit provides components that access sensors available on most personal mobile devices. This includes physical sensors such as accelerometers, microphones, magnetometers, GPS as well as Wi-Fi and Bluetooth scanning. In addition, the toolkit encompasses components that provide access to virtual sensors, for instance, personal calendars.

For preprocessing, the toolkit contains various components for signal processing and statistical analysis. This includes simple components that compute averages, percentiles, variances, entropies, etc. over data frames as well as more complicated components such as finite impulse response filters, fast Fourier transformations, gates, etc. Furthermore, the toolkit also contains a number of specialized feature extraction components that compute features for different types of sensors such as the spectral rolloff and entropy or zero crossing rate, which are used in audio recognition applications [LPL+09] or Wi-Fi fingerprints, which can be used for indoor localization.

At the classification layer, the toolkit contains a number of trained classifiers, which we created as part of the audio and motion recognition applications. Finally, there are a number of platform-specific components which are used to forward context to an application which enables the development of platform-independent classifiers. On Android, for example, a developer can attach the output of a classifier to a broadcast component which sends results to interested applications using broadcast intents. We have also developed a number of components that are useful for application development and performance evaluation. These includes components that record raw data streams coming from sensors as well as pseudo sensors that generate readings using pre-recorded data streams. Together, these components can greatly simplify the application development process on mobile devices as they enable the emulation of sensors that might not be available on a development machine.

3.2.2 Activation System

To fully understand the context of a person, it is usually necessary to recognize more than one context characteristic. As an example, consider that to know if a person is working in his office, context characteristics such as his location, pattern of movement, types of meetings and classification of ambient sounds are required. As described earlier, such context characteristics can be detected using the component system by developing configurations with the appropriate components, parameterizations and connections. Furthermore, in order to fully identify a particular context, more than one configuration would be needed at a particular time. In real life, however, the context of an entity does not remain static and over the period of time, it requires detection of different context characteristics.

Moreover, the context of a person depends on the task that the person is involved in. In other words, to know the context of a person, it is essential to know the current task. Furthermore, these tasks often follow certain patterns, e.g. tasks that a working person usually has consist of waking up in the morning, dressing up according to the weather, traveling to the work place, sitting in the office, holding meetings and discussions, going for lunch and coffee breaks, working on a computer, going for shopping, going home, relaxing, having dinner, sleeping, etc. Thus, the resulting routine is often predictable, at least partially.

Given the presence of such regular patterns of reoccurring tasks, the goal of the activation system is to exploit the knowledge about their existence in order to minimize the amount of sampling and processing that is needed to detect the user's context. To do this, the activation system enables the developer to model individual tasks as a set of states that occur sequentially. For each of the states, the developer may specify a set of configurations that describe the context that shall be recognized. In addition, the developer specifies a set of transitions between the states that define possible sequences. Using this model, the activation system takes care of executing the right configurations at the right time as shown in Figure 3.6. In the following, we describe this basic idea in more detail.

3.2.2.1 Activation Model

In the GAMBAS data acquisition framework, the modeling of the routines of a task is supported by the activation system, which uses a state machine as its primary model. Specifically, the activation system enables the automatic, state-based activation of different configurations associated with developer-defined tasks. Hence, in the activation system, the entity's context is modeled as a state with different configurations associated with it, irrespective of its granularity. The transitions between the states are modeled using context-dependent rules. In the following, we discuss these concepts in more detail.

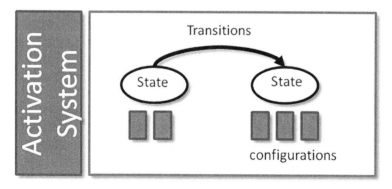

Figure 3.6   Activation System Overview.

3.2.2.1.1 *States*

A state refers to a particular decision point during the execution of a larger task. It entails a set of configurations that individually detect different context characteristics but collectively identify one of the possible decisions taken by the user.

For this purpose, states may be used to model decision points at different levels of granularity. An example of a coarse-grained state is shown in Figure 3.7(a). In this example, a high-level "working" state may encompass configurations that detect whether the person is in a meeting, working in his office or having lunch at the canteen. An example for a fine-grained use of state is shown in Figure 3.7(b). Here, the state "Fast Sampling" may be used in conjunction with a "Slow Sampling" state in order to control the precision of a certain set of configurations such as a movement detector or a sound classifier.

3.2.2.1.2 *Transitions*

Transitions are defined by the conditional changes in the configurations associated with a state. When the changes suggest that a certain condition holds, the activation systems disables the current state and its associated

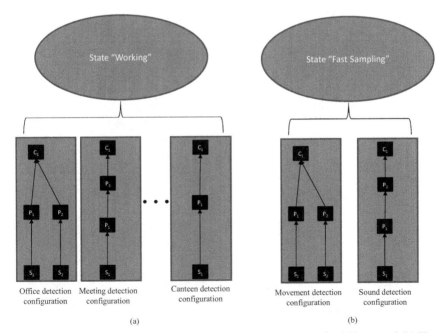

Figure 3.7 Examples of Activation System States. (a) Coarse-grained Usage and (b) Fine-grained Usage.

configurations and enables the ones associated with the new state. The activation system uses rules to model the conditions. Internally, each rule is represented by an abstract syntax tree, in which expressions for each configuration are defined. Depending on the evaluation of the expressions, the activation system decides whether a state must be changed.

Figure 3.8(a) shows two example states. State 1 has two configurations, Configuration A and Configuration B. State 2 also has two configurations, Configuration C and Configuration D. The transition from State 1 to State 2 is labeled as Transition 1 → 2, and the transition from State 2 to State 1 is labeled as Transition 2 → 1.

The abstract syntax tree of the rule for Transition 1 → 2 and Transition 2 → 1 is shown in Figure 3.8(b) and Figure 3.8(c), respectively. Assuming that State 1 is currently the active state, the activation system continuously evaluates the rules defined by the expression of Transition 1 → 2 and when the outcome of the expression, here represented by an AND operator, is true, it will disable Configuration A and Configuration B and enable Configuration C and Configuration D. Similarly, when State 2 is the current state, the activation system evaluates the rules associated with

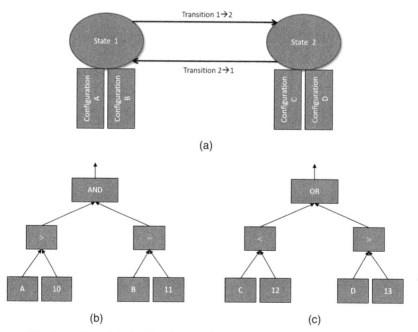

Figure 3.8 Examples of Activation System Transitions. (a) Activation System Transition Example, (b) Transition from 1 to 2 and (c) Transition from 2 to 1.

Transition $2 \to 1$ and it will execute the associated state change whenever this is implied by the outcome.

3.2.2.2 Runtime System

The main task of the runtime system is to load and execute the state machines defined by different applications. For this, the system instantiates the configurations associated with states, identifies the current state, instantiates rules for different transitions and evaluates the expressions associated with the respective transitions. Thereby, the activation system executes the state machines in an energy-efficient manner by applying configuration folding among all configurations across all the different states. The outcome of such a "folded" state machine is a single-folded configuration. Clearly, it is possible that in such a folded configuration, different configurations share the same graph structure, at least to a certain level. Therefore, the activation system provides logic for evaluating transition between the states.

3.2.2.2.1 *System Structure*

The main structural elements of the activation system are shown in Figure 3.9. These include a state machine store, the configuration folding algorithm, a rule engine and the state machine manager. The state machine store is used to

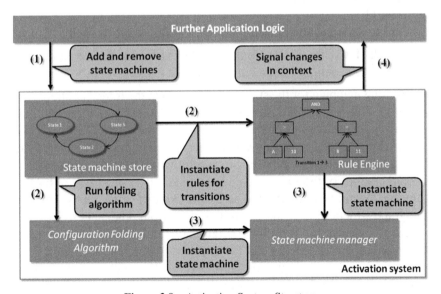

Figure 3.9   Activation System Structure.

cache the state machines associated with the applications. The configuration folding algorithm is used to compute an energy-efficient configuration for an entire state machine. To do this, the activation system applies configuration folding on the configurations of the currently executed state machines. The transitions between the states are modeled as if-else conditions and are managed by the rule engine. Once the folded configuration of the state machine and the rules for the state transitions are loaded, the state machine manager attaches the rules in the folded configuration, instantiates it and executes it. Similar to the component system, when the application logic indicates that no further context information is needed, the activation system stops executing the state machine.

3.2.2.2.2 *Configuration Mapping*

To provide a better understanding of the integration between the component system and the activation system, we describe how the configurations related to different states are folded and how the rule engine applies rules representing transitions between the states. To understand the mapping, consider an example of a state machine with two states as shown in Figure 3.10(a). Each state has two configurations attached to it. When the activation system loads the state machine, it applies the configuration folding algorithm on all configurations associated with both states, and the result is shown in Figure 3.10(b).

Let us assume that the rules for the two transitions are defined as follows:

- 1 → 2: **IF** Config. A **OR** Config. B **EQUALS** false **THEN** State 2
- 2 → 1: **IF** Config. C **OR** Config. D **EQUALS** false **THEN** State 1

The resulting mapping for the states, transitions and the folded configurations of State 1 and State 2 are shown in Figure 3.11(a) and Figure 3.11(b), respectively. If the state machine is residing in State 1 (c.f. Figure 3.11(a)), the configurations that must be evaluated according to the definition are Configuration A and Configuration B. Since folding has already taken place for all configurations of the state machine, the required graph structure for Configurations A and B is distributed across in two different graphs. However, these graph structures also share configurations from other states. Therefore, in order to evaluate the relevant configurations only, the activation system enables only the components that are required to compute Configuration A and Configuration B as shown in Figure 3.11(a). The remaining components are disabled. During the execution of the components required for State 1,

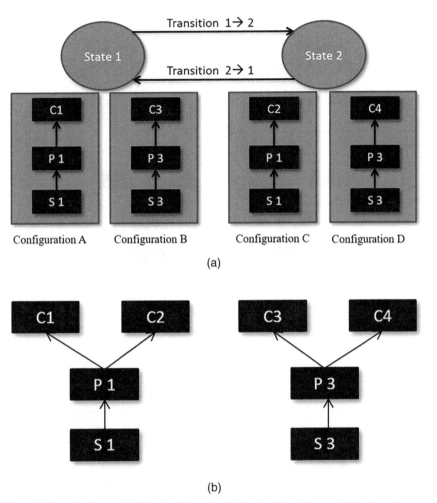

Figure 3.10 Configuration Mapping Example. (a) States, Transitions and Configurations and (b) Resulting Folded Configurations.

the activation system continuously evaluates the rule for the transition from State 1 to State 2 using the rule's syntax tree.

When the conditions defined by one of the active rules hold, the activation system initiates the state transition. Thereby, it stops the configurations of the previous state that are no longer needed and it starts the configurations required by the new state. In addition, the system stops the evaluation of the rules associated with the previous state and begins with the evaluation of the

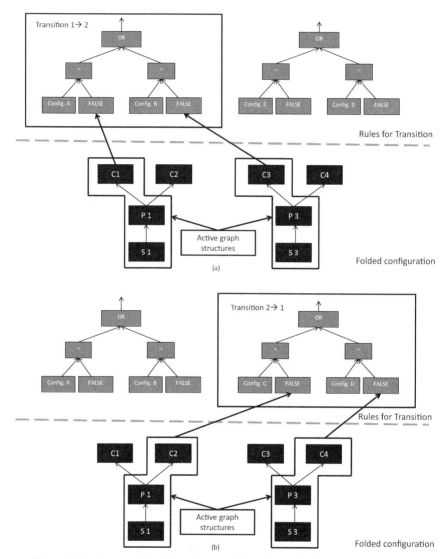

Figure 3.11   Executed Configurations and Transitions. (a) State 1 and (b) State 2.

rules for the new state. The result after transitioning from State 1 to State 2 is shown in Figure 3.11(b). Once State 2 becomes active, the system activates the Configurations C and D which are associated with State 2 and it begins the evaluation of the transition rule from State 2 to State 1.

3.2.2.2.3 *Platform Support*

Similar to the component system described previously, the core abstractions of the activation systems have been implemented using Java 1.5. In order to support multiple platforms, different wrappers have been implemented that simplify the usage of the activation system on platforms including Windows, Linux and Android.

3.2.2.2.4 *Tool Support*

Just like the component system, the activation system also provides a suite of offline tools to support rapid prototyping. These tools include a visual editor which simplifies the definition of states and transitions. The visual editor provides a user-friendly interface which allows developers to drag, drop, parameterize and wire existing configurations to create new state machines or to update existing ones. Similar to the visual editor of the component system, the visual editor for the activation system is also implemented as a plug-in for the widely used Eclipse IDE.

In addition to the visual editor, the activation system provides a set of configurations as part of the configuration toolkit for detecting different context such as location, speech, motion, etc. With the availability of the toolkit, developers do not have to create configurations from scratch. Instead, they can reuse existing configurations with trained classifiers, which can significantly reduce the application development time. A screenshot of the tool support for component system is shown in Figure 3.12.

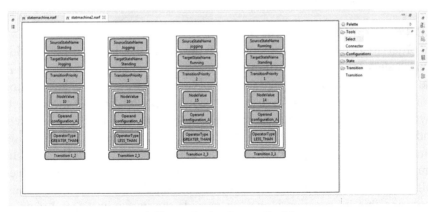

Figure 3.12 Activation System Tool Support.

3.3 Data Acquisition Components

As indicated by the previous discussions, the data acquisition framework of the GAMBAS middleware is highly configurable and extensible to support the acquisition and processing of arbitrary data from different sources. Using component compositions and state-machine definitions, even complex context recognition tasks can be supported in a highly structured manner. In order to speed up the development of applications, the data acquisition framework contains a set of basic recognition stacks including (trained) classifiers that support a broad variety of low-level and high-level context acquisition tasks. Using these building blocks, we have realized a broad number of applications described in more detail in Chapter 6. However, since they are usable beyond the scope of these applications, we briefly describe them in the following.

3.3.1 Context Recognition

The context recognition components are the basic building blocks of a context recognition application. The component toolkit provided with the component system consists of a large number of sampling, preprocessing and classification components. These components can be used to create new applications. Moreover, with the help of the toolkit, developers can implement their own components with little effort. Due to the targeted application scenarios described in Section 1.3, the components that we developed with the GAMBAS middlware are primarily focusing on location recognition, trip recognition and sound recognition.

3.3.1.1 Location Recognition

In order to determine the location of the user, the location recognition components integrate GPS information with RF signals that are present in the user's environment. Specifically, the components combine information from GPS, GSM and Wi-Fi sensors of the user's phone. Each of them has its own advantages and limitations but their collective use can provide efficient and accurate location recognition. With the widespread use of Wi-Fi, a user can typically see multiple Wi-Fi access points in the surroundings. With the limited range or signal strength of a typical Wi-Fi signal, a user can see different set of access points as he moves from one location to another. Thus, capturing this information alone can provide the user with a good view of his location. However, in places where Wi-Fi signals are not available or are very weak, GSM signals can be used instead. Typically a mobile phone can report up to 6 neighboring cell towers. Though the range of a GSM cell tower is usually large and same locations may exhibit identical cell information, together with Wi-Fi, GSM can provide accurate location information as well. Lastly, GPS signals are used to identify outdoor locations where Wi-Fi and GSM signals are not present or unique. Since each of these technologies have different energy requirements, they are used in a staged fashion that allows a user to run the location recognition continuously without draining the phone's battery.

3.3.1.2 Trip Recognition

The location of a user is an important piece of information for both users and service providers. Similarly, having information about the mode of locomotion between two locations can be beneficial for service providers. Knowing how trip was done – i.e. whether the user went on foot, took a car or a bus, stood in the bus or was able to find a seat – can help public transit providers to offer better services. In order to determine the mode of locomotion, the GAMBAS middleware encompasses multistage classifiers which integrate different sensors including accelerometers, Wi-Fi scans and GSM cell-IDs. Thereby, the classifiers use accelerometer samples to identify the general motion of the user. This allows them to determine if the user is walking, running, climbing stairs, etc. If a continuous detection of walking or running is detected between the locations, they can derive that the user was traveling on foot. If the user is not walking, the trip recognition components are using Wi-Fi and GSM cell information to estimate the movement speed of the user, which can then be used to narrow down the remaining modes (e.g. driving in a car, riding a bus, etc.).

Given a suitable infrastructure, such as the one deployed in the city of Madrid, it is even possible to identify the actual vehicle type (e.g. a specific public bus running on a particular bus line). However, even if this infrastructure is not available, it is still possible to derive the movement modality with high accuracy. In order to measure the accuracy of configuration for the trip recognition, we performed a number of validation tests over the data gathered from different modes of transportation. The final classifier with the overall of accuracy of 91.4% and the confusion matrix are shown in Figure 3.13 and Figure 3.14, respectively.

These results have been gathered by capturing training data from 4 persons in Duisburg and Bonn over the course of multiple days. Consequently, there might be a bias regarding the fit for this particular area and overall the results may be worse when applied to different areas or users.

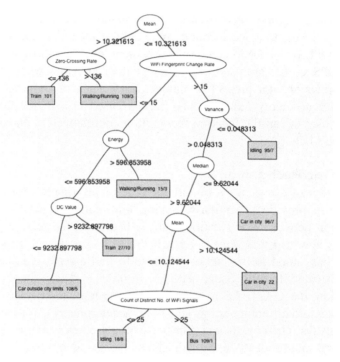

Figure 3.13   Trip Recognition Classifier.

	Bus	Walking/Running	Train	Idling	Car in city	Car outside city
Bus	108	0	3	9	3	4
Walking/Running	1	115	1	1	0	0
Train	0	0	116	4	0	3
Idling	3	0	0	98	4	0
Car in city	2	1	2	3	106	2
Car outside city limits	1	3	5	1	6	95

Figure 3.14   Trip Recognition Confusion Matrix.

However, given the high accuracy of the results, it is conceivable that this approach is broadly applicable in general.

3.3.1.3 Sound Recognition

The sound recognition components make use of audio-data collected on the mobile device and combine it with location data. They can be used for two major purposes. First, they can be used to identify features of the user's environment as done with noise recognition. Second, they can be used in

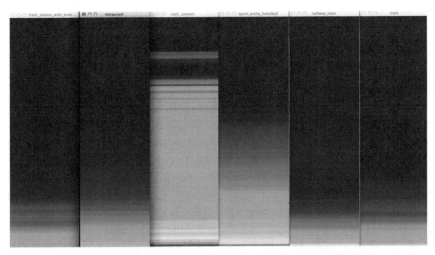

Figure 3.15 Average Frequency Vectors (Train Station, Restaurant, Rock Concert, Sport Arena, Subway, Train).

order to provide a natural way of performing user inputs as done with Voice Tagging or Voice Control.

3.3.1.3.1 *Noise Recognition*

There are several user contexts that come along with a characteristic sound environment. Being on a crowded bus, for example, a person is surrounded by a constant bus engine sound as well as human voices and other noises created by a crowd of people. This can be exploited to extract information about the user context from audio collected on the mobile device as well as to gather information about the public transport traffic situation in the whole city. To do this, we collected data sets using mobile devices carried around the city by test users. The devices are used to record several distinct audio environments like crowded bus stations and traffic jams. The collected data is then used to create sound profiles of different environments, e.g. crowded, not crowded, rush hour, etc. To do this, we compute an average frequency vector from individual samples. As shown in Figure 3.15, the average frequency vectors are different depending on the characteristic sounds present in an environment.

In order to classify recordings, the noise recognition components compute the average frequency vectors of new samples and compare them with the known profiles using the Euclidean distance between the new and all known vectors. In order to minimize the number of comparisons, we use K-Means clustering to reduce the candidates to one (good) representative for each sound profile.

3.3.1.3.2 *Voice Tagging*

Every person typically has certain locations that he attends frequently, for example, his home or his work place. To enable the user to enter these locations as destinations in a more efficient way, the voice-tagging component enables users to speed up repeated inputs. In a first step, it allows the user to add a short audio tag to his current whereabouts. For this purpose, an application typically offers a button saying "voice tagging", which, when pressed, starts a short audio recording. Typically the audio input will contain a sequence of one to three words spoken by the user. This audio is then stored in the database in a reduced form, together with the current geo data, provided by the location recognition component. At any later point in time, the user can refer to his audio tag by speaking the words used for the tag again. So if he had tagged a place by saying "my favorite restaurant", he would just have to phrase these words again to select the tagged location. At first sight, this component looks like a speech-recognition-application. However, the required computations to perform the matching between the stored voice tags and the user input are much simpler. In addition, the integration of voice tags into an application is also easier, since it does not require the definition of a grammar that defines the possible inputs. However, in contrast to voice control, voice tagging requires more effort on the side of the user, since the user has to set up tags in advance to be able to use his or her voice as an input.

3.3.1.3.3 *Voice Control*

The idea of the voice-control component is to enable the user to tell the application where he wants to go next by simple speech input. Typically, the component is activated via a voice-control button in the user interface. Once the button is pressed, it will start to receive audio data and return the recognized location. A typical speech input would be "I want to go to the main station." To enable voice control, we have integrated a customized version of the Sphinx speech recognition engine for which we developed custom models to support different target languages (including German, Spanish and English). In addition, we have developed a custom grammar for the applications described in Chapter 6. Due to the specific application scenario, the grammar includes a general list of public transport stations in the city of Madrid and it contains template sentences that are frequently used to specify routing targets such as "how do I get to Moncloa" or "compute a route to Atocha".

3.3.2 Intent Recognition

The intent recognition components take the recognized locations and trips and provide future predictions on them. Knowing the current location and mode of user transport provides significant opportunities to the service providers to improve their business, but the added ability to predict how long the user will stay at a particular place and what would be his next destination could help service providers to offer even more useful services. Apart from the service providers, a user can have many personal applications that can take advantage of this information. For instance, there can be a device charge reminder application which can alert the user to charge the batteries, based on the predicted duration of his stay at the current location and also his intended next destination. With respect to intent recognition, the acquisition framework provides duration prediction and destination prediction components.

3.3.2.1 Duration Prediction

Knowing how a long a user will stay in a particular place requires storing user location and running an offline analysis to compute predictions for the duration of user's stay in the same place in the future. There can be different options to store information about user's stay in a particular location, e.g. this information can be stored in users' device, in the cloud and also at third-party trusted servers. Clearly, storing such information elsewhere than on the user's device is prone to privacy issues and thus for the scope of GAMBAS, this information is only stored on the user's device. During the training phase, whenever a user visits a new place or a place that he has visited before, the duration prediction component stores how long the user stayed there, at which day of the week and at, which time of the day. The system then performs offline analysis on this data in addition to previously stored data which includes the information about the frequency of the user's visit to that location and his usual next locations. The system then runs a prediction algorithm to compute new predictions or update existing ones. In order to minimize the impact on the battery of the user's device, the offline analysis of data is usually done whenever the device is plugged to a socket and charged for a longer time, e.g. during the night.

3.3.2.2 Destination Prediction

In addition to knowing the current location and stay of a user in a particular location, the ability to predict the next user destination is also very useful. This information can be used to compute the transport routes proactively, for example. Similar to the duration prediction, the destination prediction is performed by analyzing the history of places visited by the user. This mainly involves identifying some sequence in the places visited by the user, the time of the day, the day of the week, frequency of visits, etc. For example, we can predict that every Saturday the user first goes shopping, then goes to a fitness club and afterwards meets friends and family. Similar to the duration prediction, this information is stored on the device and offline analysis (when the device is being charged) is performed to compute new predictions or update the existing ones.

3.3.2.3 Prediction Algorithm

The prediction algorithm uses three prediction techniques, namely time series prediction, least k history predictor and a location-dependent Markov model. The time series prediction works by taking into account the history of visits by a user to a particular location. Each visit to a particular location is saved and marked by the starting time and the duration of stay at that location. In order to predict the starting time when the user is likely to visit that location again, we choose latest last m values of starting times from the history of visits. We than identify subsets of m values of starting times in the history of visits and identify sets that are close to the latest last m values. The predicted value for next user visit to that location is obtained by averaging the next starting time value following the sets of m values. At the end of this exercise, we have a set of predicted starting time of all the locations that the user might visit. In order to select a unique next location, we check whether the predicted starting time of a location is under some time threshold T. If we can find such a starting time, we select the associated location to be the next possible location. If more than one predicted locations satisfy the criteria, we choose one randomly. A similar approach is also used for determining the duration of stay. In our tests with multiple users, the prediction techniques typically range around 20–40% accuracy, depending on the regularity of the movement patterns of the user.

4

Data Processing

This chapter describes the data processing supported by the GAMBAS middleware. Towards this end, the chapter first describes the formalisms and ontologies for the data and query models. The formalisms and ontologies provide a unified view of the heterogeneous data produced by the different players in the targeted applications. Such a unified view, based on semantic descriptions of the data and the data sources, is in line with the linked data paradigm, and it not only facilitates data understanding, but also improves data discovery and integration between both objects and persons, and other sources of data that follows the same paradigm, such as the Web of Data. Based on the data and query models, the chapter introduces the general data discovery mechanisms that are used to make data available to others. Finally, the chapter describes the architecture and implementation of the distributed data storage and processing system that allows devices to cooperate with each other in a seamless and interoperable way.

4.1 Focus and Contribution

The data representation and the associated query processing infrastructure are key to allow data interoperability between the devices and services targeted by the GAMBAS middleware. This is particularly important given that behavior-driven services often base their decision on data coming from multiple sources. Descriptions of the data and the data sources should be available to all devices. Such descriptions can include the features of interest, accuracy, measuring condition, time point, location, etc., and they are essential for search and discovery when an Internet-connected object is confronted with a large number of data sources. The query processing needs to account for the dynamic nature of some of the generated data, and it should be done in a distributed fashion, whenever possible, to improve scalability and also to increase privacy-level of data processing.

4.1.1 Data Representation

There have been a lot of efforts in employing Semantic Web technology to semantically enrich sensor data [WZL06], [BFL+07], [SHS08], [RMLM09], [PHS10]. In order to allow easy integration with other data sources available in a Linked Open Data (LOD) cloud, [Lin12] suggests that sensor data sources should be published following the Linked Data principles [BHBL09] – a concept that is known as Linked Stream Data [SC09]. The advantages of such an approach are manifold. Not only would it support the direct integration of sensor data with the large amounts of already available web and enterprise data, but it can also benefit from a large body of work and infrastructure from existing research areas such as LOD, Web and Data Base Management Systems (DBMS). One example scenario is the case where GPS locations streamed as Linked Data are combined in real time with a Cocitation Collection Service available in the LOD cloud. The service can then notify an author if there is any other author in the same location whose papers he cites. However, the state of the art in Semantic Web technologies is inadequate for sensor-generated data, due to the highly dynamic and temporal aspects of this data. Moreover, the data representation suggested by Semantic Web technologies typically are not suitable for devices with limited data storage.

Stream elements of Linked Stream Data are usually represented as RDF triples with temporal annotations. A temporal annotation of an RDF triple can be an interval-based [LPSZ10] or point-based [GHV07] label. An interval-based label is a pair of timestamps, which commonly are natural numbers representing logical time. The pair of timestamps, [start, end], is used to specify the interval that the RDF triple is valid. The point-based label is a single natural number representing the time point that the triple was recorded or received. Both approaches have their advantages and disadvantages. The point-based label looks redundant and less efficient in comparison to the interval-based one. Furthermore, the interval-based label is more expressive than the point-based label because the latter is a special case of the former, i.e. when start = end. However, a point-based label is more practical for streaming data sources where triples are generated unexpectedly and instantaneously.

4.1.2 Query Processing

The state of the art in query processing of Semantic Web data can also not be directly applied to the context of data generated by smart mobile devices. There has been work on extending Semantic Web technologies for stream data. StreamingSPARQL [BGJ08] has rules for translating continuous queries, common in stream processing scenarios, to SPARQL algebra, the standard query processing language for Linked Data. Streaming SPARQL extends the SPARQL 1.1 query language for representing continuous queries on RDF Streams.

CSPARQL [BBCG10] combines triple stores with data stream management systems (DSMS). When a continuous query arrives, it is first split into static and dynamic parts, and both parts are executed independently and results are combined at the end. EP-SPARQL [AFRS11] translates the processing into logic programs. The execution mechanism of EP-SPARQL is based on event-driven backward chaining (EDBC) rules. EP-SPARQL queries are compiled into EDBC rules, which enable timely, event-driven and incremental detection of complex events (i.e., answers to EP-SPARQL queries). EDBC rules are logic rules and hence can be mixed with other background knowledge (i.e. domain knowledge that is used for reasoning).

CQELS (Continuous Query Evaluation over Linked Streams) provides a native and adaptive query processor for unified query processing over Linked Stream Data and Linked Data [LPDTXPH11]. The query executor is able to switch between equivalent physical query plans during the lifetime of the query. The CQELS engine employs both efficient data structures for sliding windows and triple storages, to provide high-throughput native access methods on RDF datasets and RDF data streams. Similar to other systems, the CQELS engine extends SPARQL 1.1 for continuous queries. However, it also supports updates in RDF datasets as well as variables for stream identifiers, allowing queries that continuously discover streams that contain a certain property. Despite the progress in Linked Stream Data processing, currently none of the approaches consider a distributed solution for resource-constrained devices.

4.1.3 Contribution

Data representation and query processing of Linked Stream Data is an active research area with many open challenges. The GAMBAS middleware addresses the problem of data interoperability among dynamic heterogeneous data sources, where data storage is limited. It provides an infrastructure supporting the discovery of dynamic linked data sources that runs on resource-constrained devices. Thereby, it provides solutions for important aspects of continuous query processing over heterogeneous Internet-connected objects to create a scalable system that can react to changes in the network and in the data being produced.

Data interoperability is achieved by means of a unified representation of the heterogeneous data and their data sources, following the Linked Open Data principles. The unified view consists of basic vocabularies and ontologies that cover all aspects of the data required to realize the application scenarios. Special care is taken to represent dynamic and temporal aspects. The goal is to enable the devices themselves to store their generated data locally in the form of Linked Data, by using the vocabularies and ontologies provided as part of the middleware. Therefore, special care is taken to limit the amount of data that needs to be stored, since storage in connected objects is limited. To do this, the descriptions applied by GAMBAS are complete, yet compact.

To allow data discovery, the infrastructure constructs and maintains a directory of descriptions, which are accessible to every device and are constantly updated to incorporate changes in the network, whilst respecting the communication cost for each device. The directory complies with the privacy rules, by having the devices to publish only information they wish to make it public and by supporting the encryption of metadata.

To support both data interoperability and discovery, the data processing framework of GAMBAS provides Linked Data storage capabilities for all connected objects. This improves scalability and also privacy, since each device can take on the responsibility of storing its own data and it can therefore decide which data can be disclosed to which devices. There are many Linked Data storage frameworks available but none of them are designed for resource-constrained devices. The GAMBAS middleware encompasses a data storage framework based on the state of the art approaches that also complies with limitations imposed in terms of memory, processing power, battery life, etc. On top of the data storages, a query processing framework is offered that follows the same guidelines. Even though the query processing capability at each device is limited, distributed query processing techniques are integrated in order to provide a more powerful processing framework among the devices.

4.2 Data Model

As basis for interoperable distributed data processing, this section introduces the data definitions and query specifications integrated into the GAMBAS middleware. The data definition is based on an ontology that has been developed with the goal of supporting the internal mechanisms of the middleware as well as the application scenarios targeted by the middleware. The ontology and query examples are described using free text descriptions and UML-like diagrams to clarify ontological relationships among concepts and groups of concepts. These diagrams are used to facilitate the comprehension of ontological concepts and their relationships. Along with that, example instances are used to illustrate how to populate ontology instances in RDF/Turtle [W3C12d]. For the description of the example queries, GAMBAS uses a subset of SPARQL query semantics and syntaxes rather than creating a new query language. In order to enable the processing of streams of data, GAMBAS leverages the CQELS query language.

4.2.1 Data Definition

Figure 4.1 shows the GAMBAS ontologies, its classes, the dependencies among the classes as well as the external ontologies from which the ontology extends concepts and properties. The external ontologies include PIMO [Sem12], SPT [SPI12], GoodRelations[Goo12] , Ordered List[Ord12] and Vehicle Sales [Mar12]. The PIMO Ontology provides a vocabulary for describing calendaring data (events, tasks, meetings). The SPITFIRE Ontology (SPT), developed within the SPITFIRE project, aligns already existing vocabularies – such as DOLCE [CNR12], WGS84 [W3C12f] and FOAF [FOA12] – to enable the semantic description of not only sensor measurements and sensor metadata, but also the context surrounding them. In particular, the activities sensed by sensors are modeled and related with social domain vocabularies and complex event descriptions. The GoodRelations ontology is widely used to describe business and product offerings. We take advantage of the Ordered List Ontology to represent a sequence of steps. An OrderedList is a list of slots with indexes to each slot and pointers to the next and the previous slot. The Vehicle Sales ontology is a web vocabulary for describing cars, boats, bikes and other vehicles for e-commerce, and it is useful in the context of GAMBAS to generalize the means of transport of a user.

The GAMBAS ontology consists of a number of sub-classes, the generic classes being **User**, **Place** and **Activity**. In addition, the ontology contains the classes **Journey**, **TravelMode** and **Bus** that are motivated by the mobility scenario as well as **Jogging** and **Shopping** that are motivated by the environmental application scenario. In the following, we describe these classes in more detail.

4.2.1.1 User Class

The User class is used to describe users of the GAMBAS middleware. In GAMBAS, users play the roles of both data consumer and provider. As a

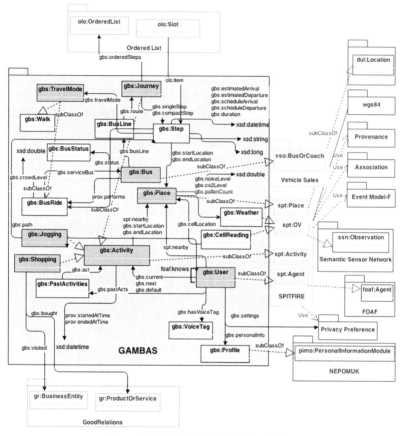

Figure 4.1   The GAMBAS Ontologies.

consumer, a user is accessing services provided through some user interface such as suggestions of bus routes or jogging areas. As a data provider, users allow GAMBAS to acquire personal data such as location and activities (e.g. traveling in a public transport, jogging, shopping, etc).

Figure 4.2 shows the User class in the GAMBAS ontology. The user class is a subclass of the spt:Agent class from the SPITFIRE ontology, which allows us to describe the user's profile such as name, email and addresses. Privacy settings are crucial in GAMBAS. To model them, we rely on the Privacy Preference vocabulary given by the Privacy Preference Ontology (PPO) [DER12]. However, during the implementation of the application prototype, it became apparent that the PPO was not suitable to describe users' shared keys and permission settings, which are needed in the privacy-preserving

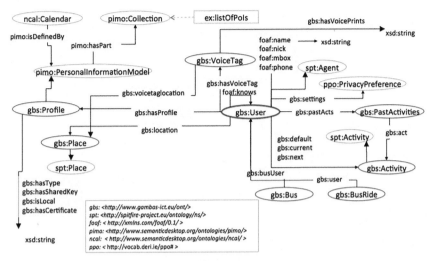

Figure 4.2 User Class.

data exchange mechanism of GAMBAS. Therefore, we added privacy-related properties to the user profile. More specifically, we extended the Profile class to include the sharedKeys and certificates used by the mechanisms described in Chapter 5.

The user's calendar information, which is used as input for the user's intent analysis, is described by creating a PIMO (Personal Information Model) instance. Users are connected to other users via the "foaf:knows" property, which allows us to list the friends of a user. The location of a user is also available and can be represented with the Place class.

Users in GAMBAS perform activities, for instance, commuting in a bus or shopping. The GAMBAS ontology provides a vocabulary to represent the user's activities, including the past, future and current ones. Past and current activities are used in combination to determine which are the user's next activities. This is done by the user's intent analysis.

Listing 4.20 shows an example of how to use the above concepts to describe a user within the GAMBAS scope, using the Turtle syntax. The example shows, among other things, how users can set access levels to other users. In this particular example, the user "John" is giving the user "Paul" access to his location. Note that the access is restricted to read-only, therefore Paul cannot modify or create instances of location for John.

To preserve the user's privacy, instances of the User class are stored in the mobile devices of the respective users. The user's location, current and next activities are dynamic properties. All remaining properties are expected to change less often and are therefore considered to be mostly static.

Listing 4.1   User Instance Example

```
ex:john a gbs:User, pimo:Agent;
foaf:nickname ``userid''^^xsd:string ;
ex:john gbs:current ex:activity1 ;
ex:john foaf:knows ex:paul ;
gbs:Profile ex:johnProfile ;
gbs:pastActs ex:archive1 ;
gbs:settings ex:ppoJohn ;
.

ex:archive1 a gbs:PastActivities ;
gbs:act ex:activity2 ;
gbs:act ex:activityn;
.

ex:activity2 a :Journey ;
prov:wasAssociatedWith ex:user ;
prov:startedAtTime ``..''^^xsd:datetime ;
prov:endedAtTime ``..''^^xsd:datetime ;
.
.

ex:johnProfile a gbs:Profile;
gbs:hasSharedKey ``B8C382391061E449CE51B29C2549BB1F'';
.

ex:ppoJohn a ppo:PrivacyPreference;
ppo:hasCondition[ ppo:classAsObject gbs:Place ];
ppo:hasAccess acl:Read;
ppo:hasAccessSpace[ ppo:hasAccessAgent ex:Paul>; ].
.

ex:activity23 a :Jogging ;
ao:mood ex:friendly ;
prov:wasAssociatedWith ex:john ;
gbs:runWith ex:paul ;
prov:startedAtTime ``2012-04-03T10:00:00Z''^^xsd:
dateTime ;
prov:endedAtTime ``2012-04-03T11:00:00Z''^^xsd:date
Time ;
gbs:path ex:runningLeg ;
.
```

4.2.1.2 Place Class

The location of a user in GAMBAS can be captured by different sensors (e.g., GPS, WIFI, GSM). The GAMBAS Place class, shown in Figure 4.3, provides different properties for the different representations. The Place class is built upon the spt:Place class, which already provides a vocabulary that includes

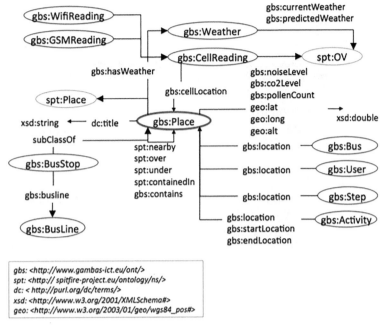

Figure 4.3 Place Class.

concepts like, city, street and GPS coordinates. The Place class extends spt:Place by enabling the representation of bus stops and cell location.

The CellReading class extends the spt:OV class, which provides the vocabulary to describe sensor observations. A noise level can be associated with every location, which can be used, in combination with the user's preferences, to suggest optimal travel routes. In addition, the place class adds properties related to the environmental scenario, such has CO2 levels and pollen count.

It is important to note that locations can be described by the set of locations it contains. This allows us to aggregate information from smaller areas, to generate a broader view. Lastly, as bus stops are a very relevant type of place in the mobility application scenario of GAMBAS, we introduce a subclass of Place, called BusStop, to specifically model them. In addition, we can have a property associated with a bus stop that lists all the bus lines that serve that stop.

A directory of locations is made available via external servers. For privacy reasons, the users' current location is dynamically stored on the mobile device.

4.2.1.3 Activity Class

A user may perform different activities, e.g. visiting a location, shopping, taking the bus or train, jogging, etc. The GAMBAS Activity class, shown in Figure 4.4, provides the properties to describe an activity. Every activity can have a start/end location and start/end time. Locations are represented as instances of the Place class. For representing the time, we use the xsd:datetime description from the OWL Time ontology [W3C12e]. Different activities, such as traveling in a bus or jogging on a particular route, are modeled as subclasses.

4.2.1.4 Journey Class

The journey class models special activities that represent general location changes of the user. A journey can involve a trip by a bus or other modes of transportations (e.g. walk between two bus stops to switch buses). A journey consists of a series of segments, or steps, and these steps are described using the class Step, which is also part of the GAMBAS ontology.

In each Step, we can specify a number of properties, such as arrival/departure times (both scheduled and estimated), duration, distance covered and start/end locations. Moreover, we can specify the travel mode used in each instance of Step, which will be described later on.

In some cases, we are interested in recording every segment between two consecutive bus stops, i.e. to check whether a user might meet a friend or not. By using the gbs:singleSteps property, we can model this case, and each Step will correspond to two consecutive points in the journey. However, we might also be interested in a more compact version of the journey, where steps in

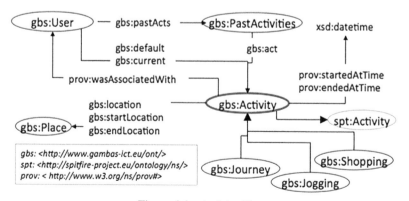

Figure 4.4   Activity Class.

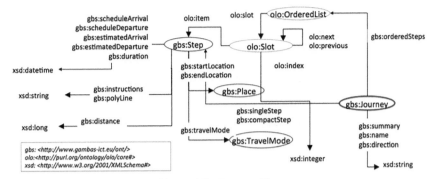

Figure 4.5 Journey Class.

which the travel model has not changed can be represented by one single step. This provides a shortcut to determine when a user entered or left a bus, for instance. For this, we have created a gbs:compactSteps property. Note that this compact version can be created at any time from the list of single steps. While it provides some redundant information, it greatly improves the performance of some queries. In addition, we also introduce a mechanism to keep track of the order in which the steps were performed during the journey. We take advantage of the Ordered List Ontology [Ord12] to represent a sequence of instances of the Step class. An OrderedList is a list of slots with indexes to each slot and pointers to the next and the previous slot. In our case, each slot contains an item of type Step. Figure 4.5 illustrates the Journey class, and an example is given in Listing 4.2.

The instances of the Journey class can be stored in the user's mobile device or a trusted external server. Information regarding the schedules is static, while the estimated departure/arrival times are usually updated dynamically.

4.2.1.5 TravelMode Class

As we mentioned in the previous section, a journey is composed of multiple steps, and each step can be performed by a different travel mode. To model this, we introduce an abstract class that represents the different travel modes. At the moment, there are two possible subclasses: BusRide and Walk, but it is straight forward to extend this by adding other means of transport, e.g. car or subway. Figure 4.6 illustrates the TravelMode class, as well as its subclasses.

For steps where a bus ride was used, we can specify further properties, like the bus used and the crowd level of the vehicle. We can also attach the

Listing 4.2 Journey Instance Example

```
ex:itinerary1 a gbs:Journey
gbs:orderedSteps ex:list1 ;
gbs:singleStep ex:step1 ;
gbs:singleStep ex:step2 ;
.

ex:list1 a olo:OrderedList ;
olo:slot ex:slot1 ;
.
ex:slot1 a :Slot
olo:item ex:step1 ;
olo:next ex:slot2 ;
.

ex:slot2 a :Slot
olo:item ex:step2 ;
.

ex:step1 a gbs:Step ;
gbs:startLocation ex:PlazaMayor ;
gbs:endLocation ex:stop2 ;
gbs:distance ''10'' ;   #distance between the two
stops.
gbs:scheduleArrival ''21:13:54Z''^^xsd:time ;
gbs:scheduleDeparture ''21:23:00Z''^^xsd:time ;.
gbs:travelmode ex:walk ;
gbs:instructions ''walk from Plaza Mayor to stop2'' ;
.
ex:step2 a :Step ;
gbs:startLocation ex:stop2 ;
gbs:endLocation ex:stop3 ;
gbs:distance ''15'' ;   #distance between the two
stops.
gbs:scheduleArrival ''21:30:00Z''^^xsd:time ;
gbs:scheduleDeparture ''21:35:00Z''^^xsd:time ;.
gbs:travelmodel ex:busride ;
```

information about the user performing the bus ride directly to this class, which can be beneficial for some types of queries.

4.2.1.6 Bus Class

A bus ride is performed by a bus, and this is also represented in the GAMBAS ontology. Figure 4.7 shows the Bus class. A bus can be associated with a stream of crowd levels to describe the number of passengers that are traveling

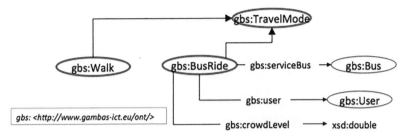

Figure 4.6   TravelMode Class.

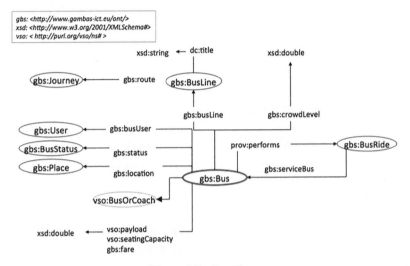

Figure 4.7   Bus Class.

on the bus. Aggregated values can be recorded and stored in instances of the BusRide class, to compute statistics of the crowd levels in the different bus routes. In addition, we can represent the route of a bus line by reusing our Journey class. Other properties include the bus line name, the bus status (in service or not) and the bus' current location.

The information about buses is provided by the transport layer and it is usually stored in an external semantic data storage. The bus location, crowd levels and its status are constantly updated.

4.2.1.7 Jogging Class
The Jogging class is a subclass of the Activity class, and it can record the path followed during the jog, the distance covered, the aggregated CO_2 and pollen

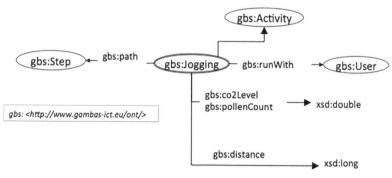

Figure 4.8 Jogging Class.

levels and the friends met during jogging. Since we do not expect changes regarding transportation mode during a Jogging activity, we can model the path taken as one single instance of the Step class, which already provides all the required properties (start/end location, polyline, duration). Figure 4.8 shows the Jogging Class.

The jogging activities are recorded in the mobile device of the user that performed the activity. However, in order to support coordination, they may be shared explicitly, e.g. with friends.

4.2.1.8 Shopping Class

In addition, the ontology includes a Shopping class, which is also a subclass of the Activity class, to describe the user's shopping. Instead of proposing a new class to model stores and their products, we use the GoodRelations ontology [Goo12], which is well known and widely used. The Shopping class allows us to enlist the products bought by the user during this activity as well as shops visited. Figure 4.9 shows the Shopping Class that are typically stored on the user's mobile device.

4.2.2 Query Specification

The data instantiated from the GAMBAS ontology is represented as RDF [W3C12a]. SPARQL [W3C12b] is the most widely used RDF query language, and therefore it has been chosen as a query language in the GAMBAS context. However, some of the data in GAMBAS is available as a stream of RDF data, or RDF streams. This is the case for the dynamic information, like the location of a user. For handling RDF streams, GAMBAS relies on an extension of the SPARQL query language, called the CQELS query language

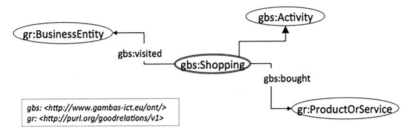

Figure 4.9 Shopping Class.

[LPDTXPH11]. The full specification of the SPARQL query semantics and syntaxes are defined by the W3C and can be found in [W3C12b]. In the RDF data model, each instance must have a globally unique URI. An RDF instance has properties that have values as literals or other instances. A literal can have text or numeric value.

In the context of GAMBAS, the SPARQL-SELECT and CQELS-SELECT queries are sufficient for all realized applications. The output of these queries is results sets in tabular form of literal and URI. Query results can be easily serialized, for example, in XML [W3C12c] or JSON [W3C13a] format. In the following, we present a number of examples for queries against the data definitions contained in the GAMBAS ontology. The main purpose of these examples is to clarify how the ontology and the definitions can be accessed using SPARQL and CQELS, respectively.

4.2.2.1 Queries on Users

For retrieving the list of all users registered at the system, we can use the query shown in Listing 4.3.

Listing 4.3 Query All Users

```
PREFIX : http://www.gambas-ict.eu/ont/
SELECT *
WHERE ?x a :User.
```

To determine the current activity of the user with a specific user identifier, we could use the query shown in Listing 4.4. Similarly, we could retrieve the user's calendar entries or friends.

For analyzing the users' intent, we can access information like the activities where a bus ride on a particular bus line was involved. Especially for the case where we want to discover whether two users have been on the same bus, we can ask for activities with a particular bus line and via a certain step.

Listing 4.4   Query Current Activity

```
PREFIX foaf: http://xmlns.com/foaf/0.1/
PREFIX : http://www.gambas-ict.eu/ont/
SELECT ?activity
WHERE ?x foaf:nick ''userid'' .
      ?activity :current ?x .
FILTER ( ?endtime > NOW ) .
```

A step, in this case, corresponds to the route between two consecutive bus stops given by the URIs of the start and end locations. In both cases, we can narrow the search to a time interval. Listing 4.5 shows an example for this.

Listing 4.5   Query Bus Rides of a Line for a Segment within an Interval

```
PREFIX foaf: http://xmlns.com/foaf/0.1/
PREFIX prov: http://www.w3.org/ns/prov#
PREFIX : http://www.gambas-ict.eu/ont/
SELECT ?busride
WHERE ?x foaf:nick ''userid'' .
      ?activity a :Journey ;
       prov:wasAssociatedWith ?x ;
      :singleStep ?step. ?step :startLocation
      <startLocURI>;
      :endLocation <endLocURI> ;
      :travelMode ?busride.
      ?busride a :BusRide ;
      :serviceBus ?bus .
      ?bus  gbs:busLine <buslineURI> .
      ?activity prov:startedAtTime ?starttime ;
      prov:endedAtTime ?endtime .
FILTER (?endtime < ''2012-04-03T00:00:00Z''^^xsd:date
Time) .
FILTER (?starttime > ''2012-04-02T00:00:00Z''^^xsd:
dateTime) .
```

The examples show that the GAMBAS ontology is flexible whether you are looking for a journey specified by start and location or other properties, such as the bus line taken. The travelMode property allows us to filter out activities where a bus was not involved.

For the user intention mining, it is important to analyze the historical information associated with buses. The query shown in Listing 4.6 retrieves all recorded bus traces of a user in a given bus.

Note that we can use the compact representation of the journey to retrieve the full segment of the user in a bus, rather than the individual steps.

Listing 4.6   Query Ride History of a User

```
PREFIX foaf: http://xmlns.com/foaf/0.1/
PREFIX prov: http://www.w3.org/ns/prov#
PREFIX : http://www.gambas-ict.eu/ont/
SELECT ?step
WHERE ?x foaf:nick ''userid'' ;
      :pastActs ?acts. ?acts :act ?journey ;
      :compactStep ?step. ?step
      :travelMode a :BusRide .
```

In the environmental domain, we can look for journeys in which some of the steps had a CO2 level above a given threshold. This is shown in Listing 4.7.

Listing 4.7   Journeys with CO2 Level above Threshold

```
PREFIX foaf: http://xmlns.com/foaf/0.1/
PREFIX prov: http://www.w3.org/ns/prov#
PREFIX : http://www.gambas-ict.eu/ont/
SELECT ?journey
WHERE ?x foaf:nick ''userid'' . ?activity a :Journey ;
      prov:wasAssociatedWith ?x ; :singleStep ?step.
      ?step :startLocation ?startLoc ;
      :endLocation ?endLoc.
      ?startLoc gbs:co2Level ?startco2.
      ?endLoc gbs:co2Level ?endco2
      OR{?startco2 > <threshold>. ?endco2 >
      <threshold>} .
```

For the above query, we need to retrieve all the start and end locations and check for their CO2 levels. We iterate over every single step on the journey to make sure we retrieve all locations visited in that journey.

Another interested query is to retrieve a list of users who had gone jogging with a particular user shown in Listing 4.8. This could be used, for instance, to indicate a stronger friendship level between the two users.

Listing 4.8   Query Users Jogging with a User

```
PREFIX foaf: http://xmlns.com/foaf/0.1/
PREFIX prov: http://www.w3.org/ns/prov#
PREFIX : http://www.gambas-ict.eu/ont/
SELECT ?user
WHERE ?x foaf:nick ''userid'' .
      ?activity a :Jogging ;
      prov:wasAssociatedWith ?x ;
      :runWith ?user.
```

As we mentioned earlier, GAMBAS extends the query set by supporting queries that involve dynamic information. For this, it uses the CQLES query language that resembles SPARQL. The main difference is the introduction of the STREAM command that allows us to specify a window of data within the stream. The query shown in Listing 4.9 retrieves the current location of a user.

Listing 4.9   Continuously Query the Latest User Location

```
PREFIX foaf: http://xmlns.com/foaf/0.1/
PREFIX prov: http://www.w3.org/ns/prov#
PREFIX : http://www.gambas-ict.eu/ont/
SELECT ?location
WHERE ?x foaf:nick ''userid'' .
STREAM <streamURI> [NOW] {?x :location ?location}.
```

In this query example, <streamURI> refers to the URI from where the stream with the data in question can be accessed. The parameter [NOW] extracts the latest location streamed. CQELS is a very flexible language, allowing an easy integration of static and dynamic data. For example, for suggesting bus stops near the user, we can write the query shown in Listing 4.10.

Listing 4.10   Continuously Query Near by Bus Stops

```
PREFIX foaf: http://xmlns.com/foaf/0.1/
PREFIX prov: http://www.w3.org/ns/prov#
PREFIX spt: http:// spitfire-project.eu/ontology/ns/
PREFIX : http://www.gambas-ict.eu/ont/
SELECT ?nearby
WHERE ?x foaf:nick ''userid'' .
STREAM <streamURI> [NOW] {?x  :location ?location}.
?nearby a :BusStop ; spt:nearby ?location.
```

It is noteworthy to highlight that CQELS queries are continuous queries, which means they are registered in the system and whenever new data is generated in the stream, the query is evaluated and results are pushed to the output. For example, we can imagine a scenario of a user walking around and getting notifications of nearby bus stops as he changes location.

4.2.2.2 Queries on Buses

This section presents a subset of queries about buses, bus stops and bus lines. For instance, to get bus stops near a particular GPS location, we can query as shown in Listing 4.11.

Listing 4.11   Query Bus Stops at GPS Location

```
PREFIX : http://www.gambas-ict.eu/ont/
PREFIX g: http://www.w3.org/2003/01/geo/wgs84_pos#
PREFIX spt: http:// spitfire-project.eu/ontology/
ns/
SELECT ?place
WHERE ?place a :BusStop ; spt:nearby ?location.
      ?location a :Place ; g:Lat ``50.0'' ; g:long
      ``3.0''.
```

Similarly, we can also retrieve the bus route for a particular bus line. The corresponding query is shown in Listing 4.12.

Listing 4.12   Query Bus Stops of a Bus Line

```
PREFIX : http://www.gambas-ict.eu/ont/
SELECT ?busroute
WHERE ? busline a :BusLine ; :route ?busRoute
```

To retrieve the list of stops covered by a bus line in the correct sequence, we can use the ordered list to iterate over the different steps as shown in Listing 4.13. Note that the query might return duplicates if start/end locations overlap. However, this can be easily fixed by a simple scan over the results list.

Listing 4.13   Query Bus Stop Sequence of a Bus Line

```
PREFIX : http://www.gambas-ict.eu/ont/
PREFIX olo: http://purl.org/ontology/olo/core#
SELECT ?start ?stop
WHERE { ?busline a :BusLine ; :route ?busRoute.
        ?busRoute :orderedSteps ?list.
        ?list olo:slot ?slot .
        ?slot olo:item ?step ; olo:index ?index .
        ?step :startLoc ?start ; :endLoc ?end
      }
ORDER BY ASC(?index).
```

With the Place ontology, we can easily query for all bus lines that run on a stop. Moreover, we can also query for bus lines that run on a given date on that stop as shown in Listing 4.14. To do this, the query looks at the routes of the bus lines and filters them by the date.

For a user waiting at a bus stop, we want to send notifications of possible delays. We can first retrieve all the bus lines that run on the stop and check their timetables against the stream of estimated times. In the query shown in Listing 4.15, we can specify a threshold (e.g., 5 minutes), and if the current

Listing 4.14   Query Bus Stop Sequence of a Bus Line

```
PREFIX : http://www.gambas-ict.eu/ont/
PREFIX prov: http://www.w3.org/ns/prov#
SELECT ?busline
WHERE <busstopURI> :busLine ?busline .
      ?busline :route ?route .
      ?route prov:startedAtTime ?start ; prov:
      endedAtTime ?end.
FILTER( ?start ><date>). FILTER (?end <<date>).
```

Listing 4.15   Query Delayed Buses

```
PREFIX foaf: http://xmlns.com/foaf/0.1/
PREFIX : http://www.gambas-ict.eu/ont/
SELECT ?estimateddeparture
WHERE ?x foaf:nick ``userid'' ; :location ?stop.
      ?stop :busline ?line .
      ?line :route ?route .
      ?route :singleStep ?step .
      ?step :startLocation ?stop ;
      :scheduleDeparture ?scheduleDeparture
STREAM <streamURI> [NOW]
      { ?step :estimatedDeparture ?estimated
      Departure }.
FILTER (?estimateddeparture >
          ?scheduleDeparturel +threshold).
```

live departure time estimation is over the threshold, then the system will notify the user.

The last query examples are related to the crowd-level information available for different public transit vehicles. To access the latest status and crowd-level information of a particular bus, we can use the query depicted in Listing 4.16.

Listing 4.16   Query Latest Crowd Level of Bus

```
PREFIX : http://www.gambas-ict.eu/ont/
SELECT ?crowdLevel ?status
WHERE ?bus a:Bus
STREAM <streamURI> [NOW] {?bus :crowdLevel ?
crowdLevel}.
STREAM <streamURI> [NOW] {?bus :status ?status}.
```

Using the GAMBAS ontology, we can store an aggregated value of crowd levels recorded for a particular step of a journey. This value can be, for instance, the maximum crowd level at any stage of that step or the average value. In the query depicted in Listing 4.17, we show how to extract the maximum crowd level of a step.

Listing 4.17 Query Latest Crowd Level of Bus

```
PREFIX : http://www.gambas-ict.eu/ont/
SELECT MAX (?crowdLevel)
WHERE   ?step a :Step ; :estimatedDeparture ?start ;
        :estimatedArrival ?end ; :travelMode ?busride .
        ?busride :serviceBus ?bus .
STREAM <streamURI> [RANGE 30min]
        {?bus:crowdLevel ?crowdLevel[timeStamp]}.
FILTER (?start < timeStamp < ?end).
```

When processing data streams, we can extract windows of data, by specifying the window parameters. In the previous queries, we used [NOW] to extract the latest value. Here, we select all the data of the last 30 minutes. Note that it is not possible to specify a start/end time interval for the window operators. Nevertheless, we can take advantage of the fact that every stream data can have a timestamp associated with it. In the case of this query, we assume that the start time did not occur before 30 minutes ago, and we select the valid crowd levels during the step in the filter condition.

4.3 Data Discovery

To enable the distributed execution of queries across multiple data stores, the query processors must be able to discover the available data stores. The GAMBAS dynamic data discovery system is responsible for providing this functionality. From an architectural perspective, it is realized as a central registry service that offers two distinct interfaces: (1) a GAMBAS-based registration interface to export metadata and search for data sources and (2) a web-based administration interface that allows to configure the discovery system, check its state and browse current registrations. The discovery system is developed using the Google Web Toolkit (GWT), a toolkit for the development of web-based client/server applications, and deployed in a Java servlet container such as Apache Tomcat. Figure 4.10 shows a screenshot of the administration interface of the discovery registry.

Besides a central registry instance for normal system operation, application developers can run their own private instances of the discovery system in their local networks. This allows using separate discovery systems for development work or prototyping and isolates the development systems from each other and the central discovery system used for normal system operation.

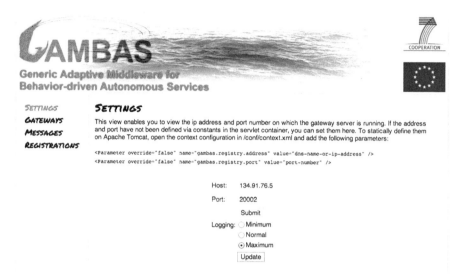

Figure 4.10   Dynamic Data Discovery Registry Administration Interface.

4.3.1 Architecture

The architecture of the GAMBAS dynamic data discovery is shown in Figure 4.11. The system is deployed as a servlet in a regular servlet container. It builds upon the GAMBAS communication system to realize remote communication and lease management as described later. The data registration is co-located with a communication gateway component that is used by the communication system to enable multi hop routing and connectivity in peer-to-peer environments with firewalls or networks with native address translation (NAT).

The co-location of the registry with the gateway allows to easily locate the registry and thus simplifies the bootstrapping of the system. The dynamic data registration contains all functionalities needed to publish metadata describing data sources, to update this information and ensure its freshness. The web-based administration interface depicted in Figure 4.10 allows to configure the discovery system (as well as the communication gateway) and to browse current metadata as well as exchanged messages.

4.3.2 Metadata Management

Metadata is used to describe data sources such that clients can easily select semantic data stores that contain data that is relevant for their queries.

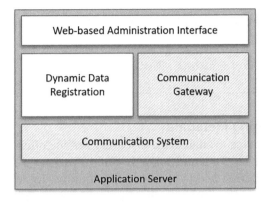

Figure 4.11 Dynamic Data Discovery Registry Architecture.

The metadata published in the registry follows the linked data paradigm to describe the data provided by devices. Listing 4.18 and Listing 4.19 show different examples of metadata information for a service providing environmental information about places and a service providing information about bus schedules, respectively.

Listing 4.18   Register a Service for Environmental Information

```
<place1> a gbs:Place ;
         gbs:noiseLevel ``-1'' ;
         gbs:co2level ``-1'' ;
         gbs:pollenCount ``-1'' .
```

Listing 4.19   Register a Service for Bus Schedules

```
<line1>  a gbs:BusLine ; gbs:route <route1> ;
         dc:title ``some_line'' .
<route1> a gbs:Journey ; gbs:singleStep <step1> .
<step1>  a gbs:Step ; gbs:startLocation <p1>;
         gbs:endLocation <p2> ;
         gbs:scheduleArrival ``00:00:00Z''^^xsd:time ;
         gbs:scheduleDeparture ``00:00:00Z''^^xsd:time .
```

It is important to note that the registry only keeps the data structure (ontologies classes and properties), but not the actual instances and property values. As the purpose of the directory is to allow discovery, it only needs to store the shape of the RDF graph, which are then used to match against user queries. An example query looking for providers of GPS coordinates is shown in Listing 4.20.

Listing 4.20   Finding Services Providing GPS Coordinates

```
SELECT distinct ?g
WHERE { GRAPH ?g
        { ?p geo:lat ?lat ; geo:lon ?lon . }
      }
```

4.3.2.1 Publishing Metadata

To make a data source available, the discovery service offers remote methods using the underlying communication system to register a new data source, to update a registration and to remove a registration. To do so, data sources send their metadata description to the registry. This metadata is then stored at the registry and made available for clients to find suitable data sources. The signature of the registration method is:

- *DeviceRegistration register(DeviceDescriptor)*

The method takes a device description that specifies the metadata to describe a data source and returns a new registration object that can be used to maintain the registration. In case a description changes, data sources can update a registration by calling the update method:

- *Boolean update(DeviceDescriptor, DeviceRegistration)*

This method takes a new descriptor as well as an existing registration (obtained by an earlier call to register) and returns a Boolean specifying if the update was successful. If the registration cannot be found in the registry, the update will fail.

4.3.2.2 Unpublishing Metadata

At some point of time, a data source might want to stop offering data or it may become unavailable. To stop offering data, a data source can deregister itself from the registry using the remove method:

- *void remove(DeviceRegistration)*

This method takes a registration and removes it from the registry. If the registration cannot be found in the registry, the method fails silently, i.e. no error notification is given. In any case, after the method finishes, the registration is no longer available for clients.

In addition to this explicit removal, the discovery service also employs a lease mechanism to ensure freshness of registrations in cases where a data source becomes unavailable without being able to deregister. To do so, the discovery service uses an existing component of the communication system.

For every registration, it starts a lease process that checks the availability of registered data sources periodically. In case a data source is not available several times, a lease manager integrated into the communication system notifies the discovery service, which eventually removes the registration.

4.3.3 Querying Data Sources

To find suitable data sources for a specific data need, clients can issue data source queries at the discovery system. To do so, they can call the find-method of the registry:

- *DeviceResult find(DeviceQuery)*

This method takes a query (implemented as an *DeviceQuery*) that specifies the intended data sources and returns a query result (implemented a *DeviceResult*) possibly including a set of suitable data sources.

4.3.4 Security and Privacy

In addition to support for public services, a secure version of the Dynamic Data Registry (DDR) provides privacy guarantees for users who may wish to limit sharing of their data to specific users or groups of users. To do this, the secure version of the registry integrates an encryption scheme known as IPHVE. This scheme not only ensures that only users with access to a particular data item are able to discover the location of the item in question, but it also ensures that the registry itself cannot become a security or privacy liability, since the registry itself also cannot read the stored metadata.

IPHVE is an attribute-based encryption scheme, which extends the Hidden Vector Encryption scheme [IP08]. IPHVE uses the Dual Pairing Vector Spaces (DPVS) framework [OT08]. Some of the main operations are:

- **Setup** – Generates a Secret Key (**SK**) and Public Key (**PK**).
- **Encryption** – Generates a Ciphertext (**Ct**) given a Message (**M**), **PK** and a Vector of Attributes (**Vx**).
- **Key Generation** – A Decryption Token (**DTk**) is generated given **SK** and another Vector of Attributes (**Vy**).
- **Decryption** – Given **Ct** and **DTk,** generates a Plain Text (**Pt**) if the **PK** used to generate **DTk** corresponds to the **SK** used to generate **Ct,** if **Vx** and **Vy** correspond to the HVE definition.
- **Test or Verification** – Returns true if, given **Ct** and **DTk**, **Vx** and **Vy** correspond to the HVE definition.

As an extension to IPHVE, a Generic Decryption Token (**GDTk**) can be generated, which allows users to set provider-defined attribute values. The **GDTk** can then be modified by the users with a Random Session Key (**RSK**), which prevents the registry to decrypt a message.

The resulting interaction with the secure DDR is shown in Figure 4.12. The message exchange remains similar, i.e. data providers publish metadata for users to discover. The novelty lies in the addition of a message from the data provider to the user with a decryption token that enables discovery. This token needs to be included in the message to the registry in order to get the results.

4.3.5 Client-side Caching

Since discovery is a mandatory step in execution of remote queries, the discovery process increases the latency experienced by applications. To mitigate this, the GAMBAS middleware provides a client-side cache that enables clients to store information about remote data providers to reuse this information in case there is another request for the same data. This is a standard approach for remote directory systems that is also used by DNS, for example. When executing a query, the mechanism first checks if it already has information about the requested data provider in a local cache. If that is the case, then this information is returned. Otherwise, a standard discovery request is issued. Freshness is provided by using standard techniques, i.e. leases and data invalidation in case of unsuccessful communication requests.

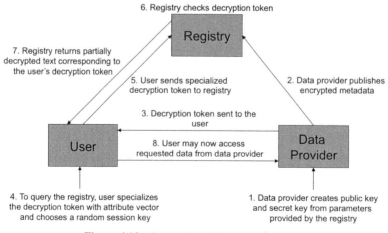

Figure 4.12   Secure Data Discovery Registry.

4.4 Data Processing

Using the Dynamic Data Discovery Registry, it is possible to discover the systems that are storing data that might be relevant for the execution of a query. However, the Data Discovery Registry only stores metadata. In order to provide security and privacy guarantees, the data itself is stored in a semantic data storage that can be queried using a query processor. In the following, we discuss the details of these two remaining components.

4.4.1 Data Storage

The semantic data storage (SDS) component provides the ability to store and retrieve RDF [W3C12a] data on devices equipped with the GAMBAS middleware. These devices range from constrained to back-end computer systems. To cope with these different device classes, two different versions of the SDS are provided: one for Android and one for J2SE environments. Both versions rely on a common (i.e. platform independent) base implementation as far as possible. To further reduce the development effort, both versions use a basic triple store for actually storing data and extend this triple store with GAMBAS-specific functionality, e.g. a remote storage interface or handling of intermittent query results (used for distributed queries).

As no established triple store exists for both J2SE and Android, we decided to use different triple stores for them and to provide a unified interface on top of them through the GAMBAS middleware. For J2SE, we use Apache Jena [Apa13], a well-established, efficient and powerful implementation that supports many additional functions such as full support for SPARQL 1.1. For Android, we use rdf-on-the-go [NUI12], a triple store implementation that is derived from Jena. On top of the triple stores, GAMBAS adds additional support for formatting query results as JSON strings according to [W3C13a]. Finally, to support formatting RDF data as N-Triple strings [W3C04], the semantic data storage contains bindings to a custom but generic N-Triple parser, called YANTRIP (Yet Another N-TRIple Parser) that is based on the JavaCC parser generator to minimize development effort and to allow for easy extensibility.

In the following, we discuss the optimization techniques applied to the semantic data storage components in order to increase their scalability. The focus of the optimizations lies on memory consumption and data indexing techniques of the storage on mobile devices. Consequently, the optimization primarily apply to the Android version of the SDS, since this version faces the most restricting constraints.

4.4.1.1 Data Storage Optimization Techniques

Reducing the memory footprint is one of the critical key targets to improve performance of the SDS [Nor07], especially when running on mobile devices. Although random access memory on mobile devices has improved, the heap size of an Android application is still limited. For example, the system RAM of an ASUS NEXUS 7 tablet is approximated 1GB, but the default memory heap size for an application running on it is only 64MB. There are a couple of reasons for this limitation. First, Android is a multi-tasking operational system with many applications stored in memory concurrently. If an application occupies too much memory, it might impact other applications or bloat the whole system. Second, Android uses the mark-sweep algorithm to perform garbage collection. Thus, an application will be paused while being garbage collected and bigger heap sizes lead to longer pause times [MNP$^+$10]. This reduces the performance of an application significantly.

To reduce memory footprint, the GAMBAS SDS for Android employs dictionary encoding which is similar to the implementations of Jena TDB or Sesame. In contrast to solutions for standard computers, we use a compact integer format that is optimized for millions rather than billions of RDF nodes. We believe this is the common scale of most mobile personal information applications. Existing RDF stores for mobile devices are restricted to smaller data sizes of approximately one order of magnitude less [ZS12]. Each RDF node is processed and mapped to a node identifier before it is loaded into main memory. A node identifier is 32 bits in size, where 9 bits are used for encoding the node type and the remaining 23 bits for encoding a string identifier. Most operations on nodes, e.g., matching during a query execution, can be performed on these node identifiers without accessing the actual string representation. Thus, only one integer must be kept in memory for each node, while string representations can be stored on flash memory. This leads to a memory footprint of just up to 12 bytes per triple, while memory profiling reported about 450 bytes per triple for the Jena memory model. Note that despite this large memory footprint reduction, we do not restrict our system to keep all triples in main memory. Instead, our RDF store can store triples in flash memory as discussed next.

For efficient access, all RDF triples are indexed with a schema we already presented in [LPPRH10]. It consists of three triple indexes with different node orders with respect to subject (S), predicate (P) and object (O): SPO, POS and OSP. The indexes are stored in flash memory to reduce the required amount of main memory and to make data persistent. We also operate a triple cache in main memory, which contains currently used parts of the indexes.

Flash memory has a great impact on the design of an efficient DBMS for mobile platforms [LNK+07]. For example, well-known B-Tree indexing techniques were shown to be not suitable for flash memory [LHLY09]. Therefore, we have built a special lightweight key-value database. This database is optimized for flash memory and allows us to fully control I/O blocking and block caching. This way we can better manage memory access and minimize the impact of Android's garbage collection due to erase-before-write limitations of flash devices [JS10].

Flash I/O is based on memory blocks. Instead of reading or writing individual bytes, the I/O unit always reads/writes a whole block. The size of a block depends on the individual devices. Thus, in order to write a single byte in a block, the whole block must be read, modified and written again. This makes random access writing very inefficient. Our aim is to reduce the number of read and write accesses as much as possible. To do so, we partition each index into individual blocks, which have the same size as the flash I/O blocks of the device. The individual blocks are stored in flash memory. A metadata structure specifies the triples contained in each block, given as lowest and highest node identifier in the sorted block. The triple cache contains a number of index blocks. If a new triple is added, it must be added to the indexes. To do so, the system loads the required index blocks into the cache. Then, the triple must be included at the right position in the index. This is trivial if the triple should be added at the end of an existing block that still has open space. Otherwise, we would need to move all triples by one position, resulting in a large number of writes. To reduce this overhead, we do not change the original block. Instead, we slice the block into two parts: an old, original block and a new one. The old one is not changed at all. The new one contains all triples starting with the newly added one. Then, the metadata structure is updated to specify that the new block contains all parts including the new triple, while the old one only contains parts before that.

As an example, imagine that a block contains three triples for subject nodes with identifiers 1, 5 and 7. The metadata will specify that this block contains triples for subjects 1 to 7. To add a triple starting with a subject node with identifier 6, we read the original block if it is not already in the cache and create a new block containing the triples starting with identifiers 6 and 7. Then, we update the metadata to specify that the old node contains triples for subjects 1 to 5, while the new one contains triples for subjects 6 to 7. We did not have to modify the original block in any way. The new block is still in the cache and hopefully will get additional triples for the same subject before writing it onto flash later. This way, we will only need to perform one write access to flash memory. To further reduce the number of read/write accesses, when we need to remove a block from the cache and write it back to flash, our strategy chooses a block that has a high chance of not being changed in the future. Together, these optimizations reduce the overhead of using flash memory considerably.

4.4.1.2 Data Storage Optimization Results

To evaluate the performance gains when applying the optimization techniques to a Semantic Data Storage, we have implemented them as part of the SDS for Android. Using this implementation, we compare the new version with the old version, which used Berkeley DB as underlying database (RDF-BDB). We also compare against the Android version of Jena TDB (TDBoid).

Figure 4.13 shows that the throughput of the improved version of the SDS (RDF-OTG) is four times higher than TDBoids and is roughly seven times higher then the original version (RDF-BDB). Moreover, besides having much better update throughput, RDF-OTG also consumes considerably less memory than other systems (see Figure 4.14). Especially, while the previous version crashed at 200,000 triples due to memory overflow error (i.e. the application consumed more than 64MB heap size), the improved version only needs 20MB heap size for the same amount of triples.

A similar trend can be seen when analyzing the response times of queries and the scalability of the optimized implementation. There, we can measure a performance increase of 20 to 200 times, depending on the query complexity. Similarly, while the original version of rdf-on-the-go was only able to handle

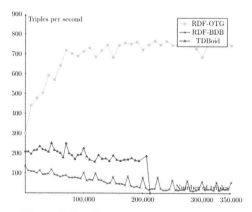

Figure 4.13   SDS Throughput Comparison.

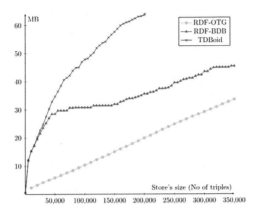

Figure 4.14 SDS Memory Comparison.

200000 triples, using the optimization techniques, it is possible to scale up to 4 million triples while still achieving response times in the order of seconds.

4.4.2 Query Processor

The query processing (QP) component enables clients to execute SPARQL [W3C12b] data queries on data sources, including queries on the local SDS, on remote SDS or on a combination of both. The GAMBAS architecture contains two different components for this: the one-time query processor (OQP) and the continuous query processor (CQP). The QP relies on the SDS to store RDF data and to execute local SPARQL queries and retrieve results for them. To enable this, the SDS provides a special interface to the QP. This interface allows direct access to the SDS in order to increase system performance. In addition to this, the interface is also used to store intermediate results of distributed queries. In the following, we discuss how the two main functional parts of the query processor, the privacy analysis and the distributed query support, are realized.

4.4.2.1 Access Control
When a query is received, the query processor has to check if this query can be executed within the specified privacy policies of the user. To do so, each query has an accompanied CallerContext to identify the sender of a query and to distinguish local and remote queries. Each query can be authorized or denied by an implementation of a so-called PrivacyManager that is described in more detail in Chapter 5. The authorization process is based on an analysis

of the received query, more specifically on the kind of data (models with data classes in the ontology) that the query will affect, e.g. a user or a location. The privacy manager can block the query, if the remote user is not allowed to access those classes. This allows a very fast authorization phase even on low-end Android devices, because it requires no result filtering. The actual privacy authorization workflow is handled by the so-called PrivacyQueryVerifier, which coordinates several internal classes to:

1. Extract the predicates for each subject in the query.
2. Derive the most probable class for each subject in the query.
3. Ask the privacy manager if the querying user might access the derived classes.

In order to support the ontology class derivation on Android devices, the ontologies are preprocessed and only an index is included inside the middleware. This avoids the overhead for parsing the ontologies, reduces the memory requirements and speeds up the analysis.

4.4.2.2 Distributed Queries

Dynamic distributed queries in GAMBAS are realized via a partial implementation of the recommendation for SPARQL 1.1 federated queries [W3C13b]. A query may contain one or more SERVICE keywords, each one specifying a sub-query on a remote data source. Following the linked data principles, data sources are identified by URIs. In principle, SPARQL 1.1 allows SERVICE sub-queries with unbound data sources. The QP does not support such queries since they can lead to a high communication overhead and may easily overwhelm restricted computer systems.

The core functionality for distributed queries consists of a generic distributed query processor and an intermediate result storage. The latter is implemented using semantic data storage. The distributed query processor receives a query and checks if it can be handled locally or contains remote parts. In the first case, it executes the query on the local SDS. In the second case, it forwards the query execution to the intermediate result storage. The result storage sends each SERVICE sub-query to the specified data source, collects intermediate results from them in the local SDS and joins them into an integrated result set. The question remains how the query issuer knows the right data sources for its query. For this, the QP uses the data discovery registry (DDR) described in Section 4.3. The general approach can be summarized as follows:

1. A query issuer wants to retrieve data from data sources of multiple remote users.
2. To do so, the query issuer first places a local query for the URIs identifying these users, e.g. based on their names or pseudonyms. Thus, the query issuer must know these users before sending them queries. Due to privacy, we do not allow users to send queries to other users that they do not know.
3. The query issuer then constructs a distributed query by adding one SERVICE sub-query for each remote user, which contains the user's URI as the URI of the remote data source.
4. This query is then placed at the QP.
5. When the QP finds SERVICE sub-queries, it accesses the local SDS and retrieves the pseudonyms of all users, whose URIs are contained in SERVICE queries.
6. With this information, the QP then contacts the DDR and requests contact information for all data sources that are bound to these pseudonyms.
7. After retrieving these, it uses this information to contact these data sources and place their SERVICE sub-query at them.

Note that to reduce the complexity for the application developers, the QP contains utilities that can be used to construct a query with all necessary SERVICE parts from a query template, in case that the remote query is identical for all receivers, e.g. querying location information for a set of users.

4.4.2.3 Continuous Queries

In addition to one-time queries, the GAMBAS data processing system also supports continuous query processing over streaming data. Similar to the one-time query processor, the continuous module also follows the Linked Data paradigm. This allows data integration among different data sources, being stream or static. Stream data is represented by Linked Data Streams [SC09], whereas the processing is supported by an instantiation of the CQELS framework for Linked Data Stream processing [LPDTXPH11].

The architecture of the module for stream processing is shown in Figure 4.15. It consists of an application client and an application server. For the full-duplex client–server communication, the system uses Websockets, which are supported by the client–server framework Netty [Net14]. In the client application for Android devices, the system uses the SDS as the Semantic Web framework, which provides an API to extract data from and write data to Linked Data Streams. The Client Publisher Handler manages

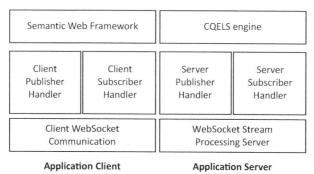

Figure 4.15   Stream Processing Module.

the upstream, which pushes RDF-triples from clients to server. To subscribe for the stream data from a particular server, the Client Subscriber Handler registers the queries to the server and manages the results listeners. Each listener listens to the results from the server through a downstream corresponding to the registered query. In the server application, the Linked Data Stream management and continuous query processor are provided by the CQELS engine. The physical streams are handled by the Server Publisher Handler and the Server Subscriber Handler. The Server Publisher Handler is tightly connected to the input manager of CQELS, in order to get the data from the clients. The Server Subscriber Handler registers the subscribed queries to the CQELS executor and routes the results to the corresponding subscribed channels.

5

Privacy Preservation

This chapter describes the automated privacy preservation framework of the GAMBAS middleware. The framework extends the adaptive data acquisition and distributed data processing frameworks to support the automated sharing of contextual information in a privacy-preserving manner. In the GAMBAS middleware, privacy preservation encompasses mechanisms and protocols to limit the access to contextual information to trustworthy clients, which also allow the user to specify which data items can be used by the system. Furthermore, it includes tools to automatically derive sharing policies by inspecting privacy settings from a configurable and extensible set of web services. Specific care is taken to avoid the use of central points of trust in order to support the policy enforcement at runtime and to maximize the applicability of derived policies to different types of context information. In the following, the chapter first clarifies the focus and contribution of privacy preservation in the GAMBAS middleware. Thereafter, it describes the privacy protocols and mechanisms and discusses the policy generation tools. Finally, the chapter presents details on the integration into the other systems.

5.1 Focus and Contribution

Context privacy is an important and very active research area in the ubiquitous computing domain. Therefore, we briefly review the state of the art in this area, before we discuss the contributions of the privacy preservation framework of the GAMBAS middleware with respect to security and privacy.

5.1.1 Trusted Computing Hardware

Hardware-based privacy approaches try to make use of current security technologies that enable trusted hardware design. This is usually based on the

Intel trusted execution technology (TXT). The TXT uses the trusted platform module (TPM) that is already built-in in many business PCs and laptops. The TXT only allows trusted (and cryptographically validated) software to run on the device. So the software itself cannot be tampered. This implies that the software engineering process is monitored closely and the privacy-preserving quality of the software can be approved [LZD08]. The drawback of such a design beside the costs is the inflexibility in hardware design. For example, simply attaching a new hardware device will tamper the security, so "secure" versions of all hardware components that are used for context processing are necessary. This includes "common" components like a USB-controller or a storage-controller. Thus, in summary, all these approaches are based on special hardware and need the complex creation of trusted software.

5.1.2 Key Exchange and Derivation

Context privacy can be achieved in several ways. One possibility is creating a common symmetric key with users and devices which are allowed to access the produced context [Mis08]. This approach is often used in eHealth scenarios. An alternative is to derive keys based on the context information in the surroundings [HV09], [RB04]. The context information used to create these keys is usually based on physical characteristics like the acoustic "fingerprint" of a room or a similar "fingerprint" based on Wi-Fi radio signals. In general, common symmetric keys cannot be used in dynamic environments, because a new key must be created and redistributed if some device leaves the group of devices that are allowed to access context information. Many of the approaches that create keys from the context information that persists in the current environment either need servers and a central authority [HV09] or can only be used in a very limited physical region [RB04]. Besides using encryption to keep the transferred context information secret, it is possible to create hashes that are distributed instead of the original context information. Because a hash is a one-way function, the original context cannot be reconstructed easily. This is similar to an approach that uses hashes and pseudonyms to hide the context from unauthorized access [EBBS07].

Another centralized approach shifts the context that can be accessed to a central database [HM08]. Similarly, it is possible to rely on several third parties that store (possibly private) context information [MMG11]. Here, the user is supposed to control the access to this context information using permissions (or a user-defined policy). When a user issues requests to a context-based service, k-anonymity [Swe02] can help to make the accessing user anonymous. However, relying on central databases always means that the user has to put trust in the database providers with regard to their compliance with the user's policy. Additionally, each provider needs to store the data securely; otherwise, a data leak may make a user's private context available.

5.1.3 Obfuscation and Generalization

Obfuscation and generalization of context information can be used to provide context privacy, usually by blurring the context. Often, these techniques are used for the privacy of location [XC09], [ACDCdVS08]. Although k-anonymity is also suggested as a solution to the privacy of location [GG03], [ZH09], [SHL+05], its use is also disputed [STD+10]. Other approaches for location privacy rely on the collaboration of users, either with [SPTH11], [RR98] or without user interaction [BS04]. MobiCrowd [SPTH11] allows users that request information from the location-based service (LBS) to share the information among each other. The information is signed by the LBS, so it can be verified by each user individually. Besides the fact that the LBS cannot gather information about users that share the context among each other, the LBS is queried less regularly, so this also has an effect on load balancing. This idea is roughly based on the use of crowds to anonymized requests [RR98]. Here, a proxy technology is used that (randomly) forwards web requests (e.g. http, ftp, gopher, etc.) to other computers or to the target server on the Internet to make the original user, who created the request, anonymous.

An approach for location privacy which is not based on the interaction between different users uses the so-called "Mix Zones" [BS04]. In Mix Zones, users are changing their pseudonyms secretly to maintain location privacy. Mix Zones require specially marked zones that cannot be used by location-based services. Additionally, using a map, many traces through mixed zones might be guessed successfully due to normal human movement behavior. The IETF working group called "Geopriv" [IET13] is also focused on location privacy. The Geopriv working group uses central servers that apply a user-specific policy and send the data to the location-based service if the policy did approve it. An evaluation of the privacy risk of location-based services [FSH12] using traces from real users concludes that current solutions which use user anonymity are effectively not providing location privacy and this result may encourage "the use of distributed solutions in which users store maps and the related information directly on their mobile devices."

The generalization of context information can also be done with other context information, especially when the information encompasses numerical values like age or height [PRAB08]. The quality of a service customized on this context information might of course be lower than if the actual context information would have been used; however, the user's privacy is still preserved.

Social networking sites often contain different context information. Additionally, they usually allow a fine-grained access control policy to be defined. Helping the user in creating and maintaining this policy as well as extracting policy information out of the social network will allow an in-depth analysis of privacy settings [FL10]. A similar approach is taken by the privacy policy tool PRiMMA [WCMS10] that allows editing privacy settings in social networks more fine-grained than supported by the network itself. The tool allows co-ownership of shared data (e.g. photos) and allows all owners to edit the privacy settings. Since social networks currently do not support a more complex privacy policy, it is necessary to store the context data on an additional server and use a separate viewer for policy editing. Often, social networks cannot be trusted with private context information, so a decentralized social network that stores the user's profile on the user's devices [NPA10] provides a solution. The necessary access control is directly enforced by the user's devices, according to the user's policy that must be specified beforehand. To have a high availability of the user's profile, the profile's context data is distributed among devices from different, trusted users. Another approach, comparable to our approach in GAMBAS, uses a server-side aggregator [JJFZ11] that crawls through different social networks and collaboration tools to retrieve the user's context that should be shared between users and devices. The user needs to specify a common profile and edit her privacy settings, defining a privacy policy. All approaches which target the privacy policies in social networks usually involve manual user actions that need to be done additionally to defining the privacy settings in the social networks.

5.1.4 Contribution

Privacy-preserved sharing of context information is a very active research area without providing the user with a clear solution. GAMBAS provides concepts and mechanisms that focus on the automated privacy-preserving sharing of context information while still being applicable for devices in the ubiquitous computing scenario, which includes heterogeneous devices, mobility and resource constraints. Hardware-based approaches for privacy-preservation need special hardware and a defined software development process that allows security audits, which define the "trust" in software. Since GAMBAS is using a dynamic architecture in software and hardware, this approach is not feasible.

Current approaches for context privacy often rely on (external) databases run by third-party providers. In contrast to that, GAMBAS does not rely on central databases, so no infrastructure is necessary to share context information in a privacy-preserving way. Key derivation for context privacy is usually only applicable in special environments and not a general solution in a pervasive scenario where devices exhibit mobility and are not bound to any infrastructure. Additionally, the necessary configuration conflicts with the goal of a distraction-free usage of devices. The context generalization in GAMBAS extends the existing approaches. If a generalization path is available that would make the context information privacy-preserving according to the used privacy policy, GAMBAS tries to use obfuscation or generalization, so customized service access is possible while preserving privacy. This obfuscation can be done in an automatic way, without distracting the user.

Approaches that extend social networks mainly focus on the privacy policies. The policies must be created manually by the users. This requires the user to learn the usually complex privacy policy language. Additionally some approaches require a central server that stores context and/or the defined policy. This requires additional server infrastructure where the user's context is stored. In GAMBAS, we use a decentralized approach where the context is usually stored on the user's device instead of a third-party server. Privacy policies can be retrieved automatically from social networks without user interaction. Also, these approaches must be extended to be applicable to not only one social networking site, but many and to other context providers like physical sensors. To create a privacy-preserved sharing of context information, GAMBAS encompasses extraction tools that gather and generalize privacy policies from a set of web services automatically as well as an associated set of mechanisms and protocols that enforce these policies at runtime.

5.2 Privacy Framework

In GAMBAS, the data acquisition and the interoperable data representation and processing mechanisms are developed to gather and distribute all possible types of data. Furthermore, it is possible to access dynamic as well as static information using one-time and continuous queries. To protect the privacy of users, the privacy framework has to limit the data acquisition and in particular the data sharing such that it respects the privacy preferences (i.e. policies) of different entities. Enforcing the desired limits is the primary task of the privacy framework.

Conceptually, the privacy preservation framework interacts with the semantic data storage (SDS) as well as the data acquisition framework (DQF) that are deployed on each personal device. In addition, the privacy framework may also be used to limit the access to information that is provided by a particular service. For this, it is also integrated into devices that are offering the services. Using a privacy policy, the privacy framework takes care of exporting sensitive data in a way that it can only be accessed by legitimate entities. The necessary privacy policy can be generated automatically by means of plug-ins that access proprietary data sources. Furthermore, depending on the user preferences, the framework can apply obfuscation in order to limit the data precision and it can also anonymize the data in order to unlink the data from a particular user. Since GAMBAS aims at supporting the use of personal mobile devices as primary sources of data, the privacy framework supports not only traditional computer systems, but also constrained computer systems as its execution platform.

5.2.1 Overview

The architecture presented in Chapter 2 describes different views on data. Regarding privacy, there are two relevant views. One is the data acquisition view in Section 2.2.1, and the other one is the processing view in Section 2.2.2. Here, we first concentrate on the data acquisition view, before we have a look at the processing view from a privacy-preserving perspective.

The data acquisition view envisions two different scenarios. The first scenario is targeting the personal acquisition of data that is used to capture the user's behavior on behalf of the user. The second scenario is targeting the collaborative acquisition of data from a large number of users that is used to improve or provide a particular service upon request of a service provider. In both scenarios, private data may be processed. Therefore, both scenarios are relevant regarding privacy.

For the first scenario, the identity of the user is important to ensure that the resulting profile can be associated with the right user. Consequently, the acquired data may be highly sensitive from a privacy perspective. For the second scenario, the user's identity is not that important, since often an aggregated view of the data will be used. Additionally, for both scenarios, it is necessary for the user to give an explicit consent to the data acquisition at least once in order to ensure that only the desired data types are acquired. To do this, the user can interact with the privacy framework by means of the intent-aware user interface to define the associated preferences. In the following, we show and describe the architectural figures from Section 2.2.1 that were extended to highlight the relevant parts for the automated privacy preservation framework.

As can be seen in Figure 5.1, the privacy preservation framework is relevant for every step of this scenario. The first two steps include the retrieval of the policy-related data from a third party (e.g. a social network or a business collaboration tool) and the generation of a personalized privacy policy from this data. A Policy Generator can create this policy using the policy language described later on in this chapter. Similarly, the third step concentrates on the policy. Here, the integration and visualization of the policy in the user interface is the focus of this step. It enables the user to manually modify the automatically generated policy to suit his or her needs. The last two steps concentrate on the data acquisition and the storing of collected data. For privacy reasons, the user may limit the data acquisition directly at the data acquisition framework, actively avoiding the gathering of certain data. A second filter step includes the short-time or long-time storage of data in the device-based registry. The user may limit or modify (e.g. obfuscate or blur) the stored data according to his or her policy. Since this scenario is focused

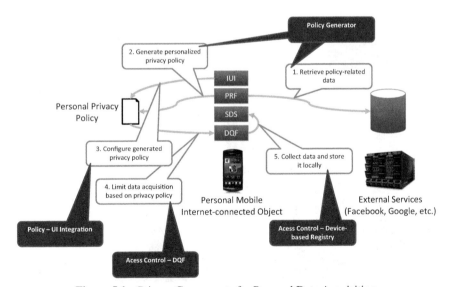

Figure 5.1   Privacy Components for Personal Data Acquisition.

on personal data acquisition, this may affect predictions that are based on the data's history, but it does not affect other devices.

The second scenario is focused on collaborative data acquisition. This includes the sharing of data with third parties, like an external SDS. The scenario is depicted in Figure 5.2. While the first four steps are identical to the ones presented for the first scenario on personal data acquisition, the last step differs. In the last step, data is not stored locally on the device, but transferred to a remote SDS where the data is stored or further processed. At this point, the privacy preservation framework needs to secure the connection to the remote service. This is done by means of mechanisms for device/service authentication and by encryption. The encryption prevents eavesdroppers from overhearing the data transmission and is necessary since the data might be transferred over insecure networks like the Internet. The authentication enables an access control component to identify the remote service and to apply the necessary limitations with regard to the acquired data. Since the data is shared with a remote service, it is often necessary to enforce a stricter policy. The access control component of the privacy preservation framework must therefore limit, anonymize, obfuscate or blur data, if requested by the policy.

In addition to data acquisition, the second relevant view is the processing view. Similar to acquisition, the processing view envisions two scenarios

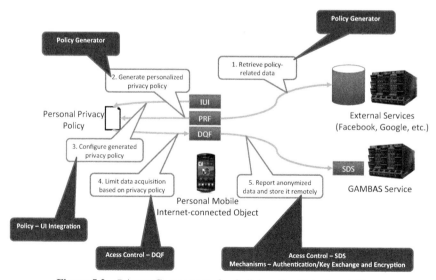

Figure 5.2   Privacy Components for Collaborative Data Acquisition.

that are relevant with respect to privacy. The first scenario describes the processing of shared data using a one-time query to the data discovery registry (DDR), and then accessing the shared data source. In the second scenario, a continuous query is executed at the continuous query processor (CQP) that retrieves and sends data on behalf of the user continuously.

The one-time processing of shared data is depicted in Figure 5.3, which shows how the privacy preservation framework integrates into the GAMBAS architecture for the processing of shared data. As a first step, if the data source's owner decides to share data through GAMBAS, the data source will be exported to the DDR. If now, as a second step, a device (i.e. the query issuer) initiates a query regarding the data source(s), it will look up the data sources at the DDR. After that, the query issuer will remotely access the data, if the user gave his consent to accessing and processing remote data. The consent is provided by means of the privacy policy. The remote data access makes use of the authentication and key exchanging mechanisms that are provided by the privacy preservation framework (Step 5). On access, the shared data sources check the status of the query issuer (i.e. check, if their policy allows data to be shared with this entity) and create a personalized view for this query issuer. In the last step, the query issuer uses the key that was exchanged in Step 5 to access and retrieve the data from the shared data sources. In this step, the communication channel is encrypted to prevent unauthorized devices from overhearing the data in transit.

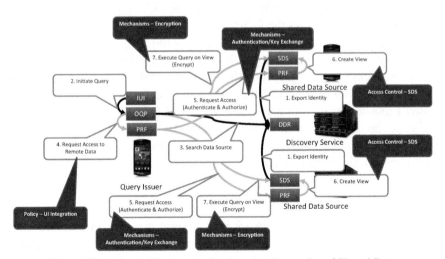

Figure 5.3 Privacy Components for One-time Processing of Shared Data.

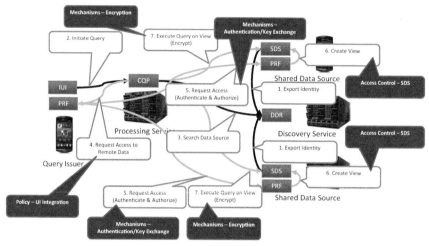

Figure 5.4   Privacy Components for Continuous Processing of Shared Data.

The second scenario is shown in Figure 5.4. In this scenario, a continuous query is executed. In contrast to one-time processing, the query is not executed by the query issuer itself. Instead, an intermediate middleware service, the continuous query processor (CQP), is used. This changes the message flow in comparison to Figure 5.3, since the CQP executes the query (i.e. performs Step 7) and not the query issuer. Therefore, the query issuer needs to trust the processing service that is running the CQP to perform queries and aggregate data reliably. As the CQP is executing the query on behalf of the query issuer, the query issuer is still requesting the access to the shared data sources. The retrieved access token is then handed over to the CQP, which uses it to execute the query. Although the message flow is more complicated, due to the addition of the CQP, the processing load decreases for the query issuer. From a privacy point of view, the CQP executes and analyzes the query, i.e. processes potentially private data. Any query issuer that is using a CQP should therefore either only request and process public data or must exhibit trust in the CQP it is using.

5.2.2 Mechanisms

The privacy preservation framework uses several mechanisms to keep data private, e.g. prevent eavesdroppers from overhearing private data, establish encrypted communication channels and authenticate users and servers. Additionally, the framework uses a privacy policy to describe which data should be shared with whom. The mechanisms then make sure that the primitives that are defined by the policy (i.e. users, groups, data and access rights) are enforced properly at any point in time. To enforce the policy with regard to users or groups, authentication is necessary. For the security of data, devices and servers need to communicate securely (i.e. using encryption communication channels). Access to the shared data is controlled by combining authentication and secure communication. Additionally, access control must be enforced depending on the different views and scenarios that are targeted by the GAMBAS middleware.

To support remote communication, the GAMBAS middleware relies on the BASE communication middlware depicted in Figure 5.5. Originally, this middleware has been developed by researchers at the Universität Stuttgart [BSGR03] and it has been refined over several years [HWS⁺10]. For example, in the European research project PECES (PErvasive Computing in Embedded Systems) [PEC12], BASE has been used to enable the secure networking of embedded devices in smart spaces over the Internet [AHM12].

As hinted in Figure 5.5, the BASE middleware provides a rather traditional object-oriented interface for the application programmer, which relies on explicitly defined service interfaces and generated proxies and skeletons. Underneath, it enables spontaneous and secure device interaction and discovery. To do this, BASE relies on an extensible plug-in model that can be used to support different communication technologies and protocols. The extensibility of BASE includes hooks for the integration of authentication and key-exchange mechanisms as well as encryption protocols. However, instead of describing BASE, in the following, we focus on the contributions of GAMBAS that are required to implement the overall system architecture. From a conceptual point of view, these contributions are independent of the

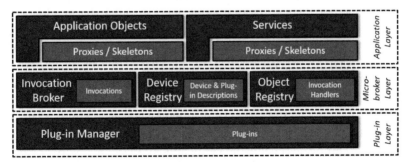

Figure 5.5 BASE Middleware.

concrete implementation and could have been implemented on top of other communication middleware systems as well. However, due to its flexible communication plug-in support, we found that implementing them with BASE was efficient.

5.2.2.1 Authentication and Key Exchange

In order to enable trustworthy and secure interaction between devices, it is necessary to authenticate interacting devices and the data exchanged between them. In particular, it is necessary to authenticate individual devices/services, e.g. during the establishment of a connection or during the access of shared data.

In GAMBAS, authentication relies on both asymmetric and symmetric cryptography, which requires the availability of keys. In the case of symmetric approaches, the keys are available only to a particular set of devices, which may use this key to ensure authenticity with respect to the devices that share the key. In the case of asymmetric approaches, the key consists of a public part (the so-called public key) and a private part (the so-called private key). The keys may then be used to authenticate individual devices.

Both symmetric and asymmetric approaches can be used to distribute further keys on the basis of existing keys. However, there needs to be at least one key available to bootstrap the overall process. Usually, this key needs to be distributed by means of a secure channel. Typically, this is done offline, e.g. as part of the device configuration. In GAMBAS, while still supporting this type of key distribution, we also offer a more convenient key exchange for user-to-user authentication, which is as secure as the underlying service.

5.2.2.1.1 *Server Authentication*

Server authentication enables the authentication of a server or a server-based service to another device. The other device can either be another server (for server-to-server communication) or a user device (e.g. a smartphone). In GAMBAS, servers are used to host services like a traffic information service or GAMBAS-related services like the CQP. The authenticity of these servers and services is important since GAMBAS applications rely on the data retrieved from them.

It is noteworthy that the server infrastructure envisioned by GAMBAS is similar to the server infrastructure in other networks, like the Internet. Here, pre-deployed certificates enable the verification of the authenticity of sites for purposes like Internet banking or e-mail retrieval. Since these mechanisms are in daily usage and have been proven effective for years, GAMBAS also relies on them. Each server in GAMBAS is therefore equipped with a certificate that is issued by the certificate authority or some trusted third party (e.g. a particular company).

Since certificates rely on asymmetric cryptography, this results in a key pair (a public and a private key) being deployed on every server. While only the server knows its private key, the public key (as part of the certificate) is shown to devices for authentication. Using a common certificate infrastructure, the public key is signed by the authority's key pair, which might then again be signed by the domain authority's key pair, leading to a certificate tree. An example certificate hierarchy tree is depicted in Figure 5.6.

As can be seen in Figure 5.6, it is not necessary for a GAMBAS application to trust a whole company. It is sufficient to trust only the parts of the company that are providing GAMBAS-related servers and services. Accessing a GAMBAS-related server will then trigger a certificate verification. It is possible to verify whether a particular certificate belongs to the GAMBAS-related sub-tree by recursively validating the certificate chain from bottom to top. To do this, the signatures must be verified one at a time. If the chain is valid and if it contains a pre-deployed GAMBAS certificate, the validated certificate belongs to the spanned part of the tree, i.e. it belongs to a valid GAMBAS server.

Similar to other infrastructures, the GAMBAS middleware makes use of the X.509 certificate standard. Among other things, this standard defines a common format for certificates, which enables the use of existing tools to generate keys and certificates offline. Specifically, it is possible to use the implementations provided by the OpenSSL library. This avoids the need for implementing key generation mechanisms and thus, it eliminates the need for providing tools that exist already.

For device authentication, GAMBAS uses an authentication based on the standard ISO authentication framework [CCI89], which can be used with the Diffie–Hellman (DH) key exchange in its original version (using

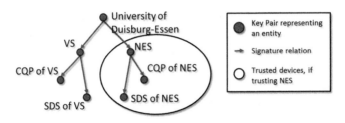

Figure 5.6 Certificate Hierarchy Example.

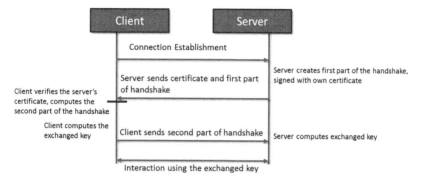

Figure 5.7   Certificate-based Key Exchange.

RSA certificates). The interaction is depicted in Figure 5.7. Additionally, the GAMBAS middleware supports a modified version of DH that relies on elliptic curve cryptography (ECC) certificates, which is more lightweight and therefore better suited for the use on smartphones or other devices with constraint resources. The exchanged keys can then be used with a key-derivation function like PBKDF2 [Kal00] to create a common shared key among any set of devices.

For pre-deployed username/password combinations, we use a hash-based authentication mechanism that does not reveal any user or password information to eavesdroppers which is important, since the past successful attacks on protocols such as MS-CHAPv2 show the necessity of a higher security standard. However, due to the focus on smartphone applications, finding the right balance between security and user convenience is a challenge.

Both the certificate-based and the username/password-based key exchanges result in the computation of a common shared key that cannot be computed by any attacker that might have overheard the communication. This key may be stored by a key store component of the middleware and used for further communication attempts, which speeds up the communication start by skipping the authorization part. This does not result in lower security, because the possession of a common shared key shows that each interaction partner was authorized properly before. Nevertheless, such a key should never become persistent. It should time-out or be renewed after a certain amount of time.

5.2.2.1.2 *User-to-User Authentication*
User-to-user authentication explicitly authenticates one user's device to another user's device, e.g. to share data between two smartphones. This can

be used to share data between users that trust each other, e.g. friends or co-workers. The authentication between users is different from the server authentication described previously, because the devices are not necessarily part of a certificate infrastructure. Only few users set up a certificate infrastructure for their private devices, so we cannot reasonably rely on user certificates.

Clearly, user-to-user authentication is not necessary in all scenarios, e.g. if a user requests information about the next bus from A to B from a service provided by the bus company. It is necessary, however, in scenarios that include the collaboration of users. This includes the sharing of data (like the current location) or a behavior profile that describes a possible future movement pattern of the user. Clearly, such private information should not be shared with anybody, but instead, it should be properly secured. As the first step to the solution, GAMBAS introduces an innovative user-to-user authentication mechanism that makes use of collaboration tools such as Google Calendar or social networks such as Facebook and that can be used as an alternative to the common infrastructure-based certificate architecture.

Many users are using social networks or similar services on a regular basis. They define trusted users in these networks by adding them to their personal network (e.g. friend relationships on Facebook). This information can be used to exchange a shared key, piggybacked on the service. To do this, GAMBAS introduces the so-called PIggybacked Key-Exchange (PIKE) [AHIM13].

PIKE can be used on any service that enables the secure restricted sharing of resources. This means that the service authenticates its users, models relationships between different users with respect to resource usage and enables the specification and enforcement of access rights. From the perspective of the users, the service performs its access control to resources properly. This means that a) it protects the resources from being accessed by illegitimate users and b) it allows access from legitimate users. Yet, beyond proper service operation, we do not assume that the service is necessarily trustworthy. Examples for these services are Facebook or Google Calendar. To use PIKE, the device of the user must be able to access the service regularly through the network. For this, the service provides some API or it uses a mobile application that synchronizes the changes to the resource.

Every time the friend relationship changes, PIKE starts to analyze the friends in order to detect new friends. In case a new friend is found, it will trigger a key exchange between the two friends, using the secure resource-sharing capabilities of the service. To do this, PIKE performs either a local modification on the triggering resource or, if this is not possible due to a limitation of the mobile application, it uses the API of the service. Once the changes have been made, PIKE simply waits for the next resource synchronization at which point the new friend will have received the key through the secure resource.

Once the personal interaction takes place, these keys can be used for authentication among the devices of the friends. To do this, PIKE simply extracts the keys from the secure resource and provides them during the interaction to the GAMBAS middleware.

To formalize this interaction, Figure 5.8 depicts the resulting logical protocol flow. Conceptually, PIKE involves three entities, namely the two devices of the interaction partners (i.e. "Friend A" and "Friend B") creating a new friend relationship and the service. To establish keys, these three entities interact with each other using three steps.

- After the change in the relationship was triggered (either through an active notification or through a regular service synchronization interval), the two friends contact the service to check if there needs to be a key established between them.
- If so, the two friends compute two keys (K_A and K_B) independent from each other and post them to a secure resource.
- In the next synchronization interval, they recognize and retrieve the key posted by the other friend. Then, they compute the combined key K_{AB} and store it on their device(s).

After the completion of these steps, the interaction partners possess the exchanged key. Once a personal interaction through GAMBAS takes place, the key (or a derived key) can be used to enable group communication as well as private communication and user-level authentication between the two friends.

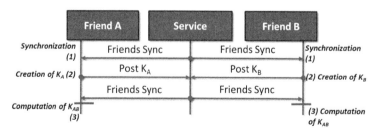

Figure 5.8   PIKE-based Key Exchange.

Figure 5.9 User-level Key Posted on Facebook.

To execute PIKE on top of the Facebook service, GAMBAS uses the Facebook Graph API to access and modify data from the social network. Each user of Facebook has a place for discussions, the so-called *wall*. This wall is used to post the keys K_A and K_B that is then automatically picked up by the friends' devices. Since friends cannot change the visibility of posts on another friend's wall, the keys are posted to the own wall. On this wall, posts can be created with a privacy setting that constraints the access to the other friend (see Figure 5.9 for an example). The friends will then retrieve their keys by going through their walls.

The combination of K_A and K_B to K_AB can use different mechanisms. While simple mechanisms like an XOR of the two values and the use of a key-derivation function to create K_AB will result in the same security as the underlying service (i.e. Facebook, which does not leak the posts, i.e. complies with its security and privacy settings), a more complex mechanism like a Diffie–Hellman key exchange can also provide security against data loss.

The key K_AB that will be exchanged after performing PIKE enables the users to authenticate each other with an exchanged key, even when their devices are not connected with the Internet, but in physical vicinity. A key for every friend relationship ensures that the authenticity is on a user-to-user basis and even malicious users cannot tamper the authentication to another user. Similar to other exchanged keys in GAMBAS, this key may be stored by the key store component and used for further communication attempts, which speeds up the communication start by skipping the authorization part. Also this key should not become persistent, but PIKE should be re-performed from time to time such that the key is renewed.

5.2.2.2 Secure Communication

Secure communication is generally used to avoid eavesdroppers from overhearing private data. In GAMBAS, the communication between different services, servers and mobile devices may contain private data. Imagine a user searching for the next bus to the mall. If this search (usually a request to a travel planner service) can be overheard, not only the next location of a user (i.e. the mall), but also the planned activity (i.e. shopping) is revealed. Similar problems occur, if personal data like audio recordings, GPS coordinates or movement patterns are shared between users. Any eavesdropper might receive this data if he is in the vicinity and can then later analyze this data, creating user profiles. To avoid this, the GAMBAS middleware relies exclusively on secure communication channels.

To establish a communication channel between two devices, the BASE middleware uses plug-ins that abstract from the used communication technology. Due to BASE's architecture, it is possible to extend this plug-in stack easily. For secure communication, we add an encryption plug-in to the set of existing plug-ins. The plug-in searches the key stored in the device local SDS for a key of the communication partner and uses this key to perform authenticated and encrypted (i.e. secure) communication. An example communication stack using the encryption plug-in is shown (for multi-hop communication) in Figure 5.10.

Although not all applications in GAMBAS require secure communication, recent publications [AHM12] have shown that the overhead by means of communication latency is small. Therefore, secure communication is activated by default and should only be deactivated for public announcements. The encryption technology used in GAMBAS is AES, a symmetric encryption mechanism, which is both fast and secure and available for all devices in the GAMBAS scenarios. AES relies on a shared key between the communication partners that must be exchanged beforehand. The authentication/key exchange in Section 5.2.2.1 shows how such a key can be established in

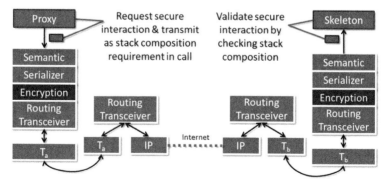

Figure 5.10   Secure Multi-hop Communication Example.

GAMBAS. To establish a secure communication, authentication is a crucial step that must not be skipped. Without authenticity, the identity of the communication partner remains unclear. If a secure communication channel does not establish identities, any data that is traveling to a (possibly) unauthorized communication partner must be regarded as public data.

After the authentication, the exchanged shared key is stored in a key store together with the device or user id. To enable secure communication, the device or user id is then used to retrieve the key from the key store. To improve the performance, the shared key can be cached after the communication for the next interaction, but should be changed regularly (e.g., by performing a re-authentication) to avoid impersonation attacks using lost keys for interactions.

5.2.2.3 Access Control

To ensure privacy, GAMBAS relies on user-specific privacy policies. Access control enforces these privacy policies. Using access control, data that is captured by the data acquisition framework (DQF) is protected from unauthorized access. In GAMBAS, access control must take into account the following three points:

- **Authentication:** A user or device must be authenticated, before it may access any resource in GAMBAS that is using access control. Therefore, it must use one of the mechanisms described in Section 5.2.2.1.
- **Encryption:** A user or device must use encryption while accessing data that is using access control. The encryption, described in Section 5.2.2.2, enforces the secrecy of the data while it is being transferred.
- **Policy Compliance:** Before any data is transferred, the access control must check the policy (see Section 5.3) for the data to be sent. The policy contains the users or devices that may access the data (if any) and the access control must follow the policy.

If these points are evaluated properly by the access control mechanisms, the policy is enforced securely. The general process of access control in GAMBAS can be seen as the execution of these six steps:

1. Device A wants to access private data on Device B. Since the data is private, Device B is using access control to protect it from unauthorized access.
2. Device A opens a communication channel to Device B. It sends the plug-in configuration for authentication/key exchange and encryption to signal the need for secure communication.

3. Using the plug-ins, the two devices authenticate to each other. Device A sees that Device B is owned by "Bob", while Device B authorizes the user "Alice" from Device A.

4. Device B now checks, if Device A is using encryption on the communication channel. If this is not the case, the interaction is terminated otherwise the interaction continues.

5. If successful, Device B checks the policy for the data that is to be retrieved by Device A. It searches for the appropriate data type and the access rights of Alice.

6. If the data type can be found and the access rights of this type allow Alice to access the data, Device B grants access and Device A can retrieve the requested data.

In general, it might not be necessary to authenticate Device B in Step 3. Nevertheless, many of the authorization schemes presented in this document are using symmetric authentication, i.e. both communication partners are authenticated at the same time. Additionally, the general process is modified depending on the communication partners in GAMBAS.

In GAMBAS, access control is used to access any private data. Since the scenarios in GAMBAS are manifold, the general access control process needs to be adapted to these scenarios. In the following, the three different access control mechanisms in GAMBAS are presented. At first, we show how access control is used in the data acquisition framework. Then, we concentrate on any device-based registry and at last, we describe how data access and access control with remote data storages is realized.

5.2.2.3.1 *Data Acquisition Framework (DQF)*

The data acquisition framework (DQF) is running directly on the user's device. It is implemented as a module of the GAMBAS middleware, which is realized as a combination of different modules. In GAMBAS, all modules are running in the same process on the device. Since processes in operating systems are isolated against each other, only other GAMBAS modules (running in the same process) can call the internal API. The PRF provides methods that allow the DQF to check whether a certain data type is allowed to be detected. The DQF must call this method before any attempt is taken, to create a recognition stack for detecting any kind of data or context. The method then returns a value that states whether the data or context is allowed to be detected or not. The DQF then changes the recognition stack accordingly to only detect the kind of data or context that is allowed to be detected by the privacy preservation policy.

This access control mechanism does not need any authentication or encryption since it limits the data acquisition directly on the device. Only GAMBAS modules can therefore retrieve and access the policy and the acquired data. On every startup of a GAMBAS application that needs to acquire data using the DQF, the DQF will check the privacy policy for any data type that needs to be detected by this application. If the policy does not allow the gathering of this data type, the application might not be started successfully, but the privacy of the user is preserved. This type of access control enhances the privacy of the user by not capturing data. Data that is not captured cannot get lost or overheard by anybody, even if the user's device gets stolen, the data cannot be revealed since it was not gathered at all. Not acquiring data is therefore a valid privacy goal that can be fulfilled in GAMBAS using the privacy preservation policy. It puts the user in the direct position of defining the data types that are allowed to be used for context or activity recognition.

5.2.2.3.2 *Device-based Registry*

Any device-based registry like the semantic data storage (SDS) stores data that was gathered by the DQF. The data is stored directly on the device itself, not involving remote interaction. Similar to the limitation of data gathering that was described in the previous subsection, this allows the access control to be performed without the need of encryption and authentication.

In GAMBAS, the data stored in a device-based registry is used to predict possible user behavior in the future. To protect his privacy, a user can choose not to store specific data on the device at all, such that no history on the device is created. Additionally, the privacy preservation framework makes it possible to mark stored data as not exportable. This can be modeled using a policy entry for this specific data type, which does not give any access rights to another user. The data is then only processed on the device itself and does not leave the device. Of course, this might result in a limited prediction since the device only has limited processing power. To mitigate this, the preservation policy that is used to limit the access is personalized to each user and may be tweaked, if it is perceived as too restrictive or too liberal.

The PRF contains a method that must be called through the API by any GAMBAS application, if data acquired by the DQF is stored on the device (e.g. using a device-based SDS). This method is similar to the one described in the previous section, but returns whether the data may be stored on the device or not. The application can then see if it is allowed to build a data history for this type of data. It must then comply with the result of this call.

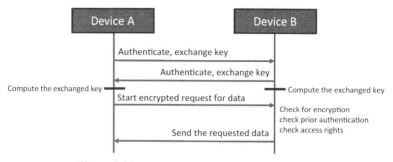

Figure 5.11   Data Request using Access Control.

In contrast to the DQF, a device-based registry like an SDS also contains a remote interface that may be called by other devices. If no data is shared, this remote interface must be inaccessible for other devices. If data is shared, the remote interface uses the device's PRF to perform access control as described in the general process of access control above. A check for authenticity and access rights, as well as the use of encrypted communication is necessary, before any private data may be shared. A simplified message flow for a successful data request can be seen in Figure 5.11.

5.2.2.3.3 *Remote Data Storage and Continuous Queries*

In some scenarios envisioned by GAMBAS, data might be stored outside the user's device. It could be stored in a remote SDS that provides additional computing capacities to give a better prediction on the future values of the data. Storing data remotely requires full user consent and could possibly breach privacy, since private data is transferred and stored on a remote device or server that is usually not owned by the user. This scenario is depicted in Figure 5.2. Here, the data is stored on a remote SDS, for example, to be aggregated for statistical purposes.

Private data is only stored remotely if the privacy policy has a valid entry for the specific private data type and it allows the sharing of the data type at this remote location. Similar to the mechanisms described before, the PRF provides a method that shows, for given values of data type and remote service or server, whether the data is allowed to be transferred there. In addition to the enforcement of the policy (which already includes the authentication of the remote service or server), the remote transfer needs to be encrypted, such that the data cannot be overheard. The access control mechanism is implemented similar to the ones described previously. Since storing data in a remote location is inconvenient for many users, the GAMBAS applications try to minimize the need for this. One important exception is the remote storing of information that users are obligated to by contract, for example, a bus company that gives out chip cards, which are validated by touching chip card readers at the bus entry, may use the travel information in an anonymized fashion, if it informs its customers accurately.

In addition to the simple one-time query that usually only needs one request–response message flow, GAMBAS supports continuous queries that may be used to notify users if the response data changes. Continuous queries do need a permanent Internet connection and may need more resources than a simple smartphone can provide (in terms of CPU power and RAM). Therefore, the continual query processer (CQP) is realized as a remote GAMBAS service.

A query involving a remote CQP changes the authorization flow, since the CQP is querying other data sources on behalf of the user. As can be seen in Figure 5.12, the device now first authenticates the remote CQP service, which will then issue a request to access a certain data source. The device must now check with the privacy policy whether the CQP service is allowed to access the requested data on behalf of the user. If the policy evaluates to true, the data source is queried to hand out an access token that enables a remote device to act on behalf of Device A. The data source will again check, if the user of Device A is allowed to retrieve the queried data. If that is the case, the data will be processed by the remote CQP service. The PRF of the data source will consult its policy to evaluate these two questions. If they evaluate to true, the data source will transfer an access token to Device A. This access token

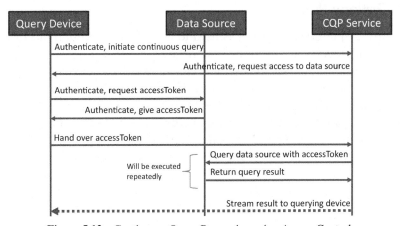

Figure 5.12 Continuous Query Processing using Access Control.

can then be used by the remote CQP service to execute the query, retrieve the data and stream the query result continuously to Device A. The CQP might execute the query repeatedly and update the continuous query result that is streamed to the device accordingly. If used with more than one data source, the CQP service needs an access token for every source and is also used to aggregate data remotely. This aggregation removes burden from the device and makes it possible to execute even complicated continual queries with resource-constrained devices.

To preserve privacy whenever possible, the remote CQP service needs to be properly authenticated and trusted by the device using it. An ideal CQP would be a home server that is in possession by the user itself. In that case, the query result will not depend on the relationship between the data source and the CQP service (since it is identical to the relationship between the source and the user's device). If an external CQP service is used, the data source could change its view on the data, since the policy on the source could have different constraints for the user and the external CQP service.

In the case of a remote CQP, all interactions between devices, data sources and the CQP must be encrypted, because they might contain private data. As shown in Figure 5.12, also all devices must be authenticated, such that the privacy preservation framework can perform the access control properly. Although the access control process is more complicated since more parties are involved, the benefits of using a remote CQP (i.e. a resource-saving query execution) outweigh the drawbacks in many scenarios.

5.3 Privacy Policy

The privacy preservation policy is used to describe the access rights of data types. Additionally, it describes how data types relate to each other. The policy is customizable for the user and can be serialized in a policy language that is based on RDF. When used in the PRF, the policy can specify which data types should be shared with which users (or companies). Therefore, the policy contains the data types and the sharing permissions, individual to each user.

The policy representation shown in Figure 5.13 displays policy permissions (i.e. using triples), which model the access rights on data types. Each

```
PermissionA affects Location
PermissionA grantedTo Bob
PermissionA obfuscation City
```

Figure 5.13 Privacy Policy Permission Example.

of these RDF triples contains a unique name as first argument, and then one of the relations "affects", "grantedTo" or "obfuscation", which denote different aspects of the policy permissions. The third argument of the triple, i.e. "affects", is the data type that should be affected by this permission. The relation "grantedTo" denotes the user that is granted this permission. Since a permission could grant the same access rights to many users, the triple using the relation "grantedTo" can occur more than once (with different users) in one permission. The user name (which could include a unique identifier) links to the profiles that this user is using on social networks or other collaboration tools. The last relation "obfuscation", optionally defines the obfuscation level for this permission. Depending on the data type, different obfuscation levels are possible. For the current location, this could be the actual GPS coordinates (i.e. no obfuscation), the current city or the current country the user is located in. Since the policy is created individually for every user, the user itself is implicitly part of every policy triple and is left out in the policy language. This means instead of creating statements like *Charlie's data type location is "grantedTo" Bob using the "obfuscation" level city*, we simplify the policy triples in Charlie's policy to the ones depicted in Figure 5.13.

The data types are specified in the data model described in Chapter 4 and used by the data acquisition framework discussed in Chapter 3. Since each different data type might provide different obfuscation levels, the levels are also defined as part of the data model. While some data types like "location" can provide more than one obfuscation level, other data types might not provide any. Thus, the use of obfuscation is optional and depends heavily on the underlying data type. In summary, both the data type and the obfuscation level are based on the data model of the GAMBAS middleware.

Another type of policy definition are triples that describe relations between different data types. These relations define a simple hierarchical relationship between data types and can be used to infer access rights for similar data or data that is used as a building block for a more complex data type. Imagine that Charlie shares his current location with Alice. If now Alice asks for the name of the street where Charlie is located, the PRF will search for the data type "street", might fail to find a policy entry for it and will deny access to it. Therefore, the policy language includes the *consistsOf*-relation. Using the privacy policy *Location consistsOf street,city,country*, the data type "street" can be found and if there are no *Permission* policy relations for "street", the "location" data type is checked. In this example, Charlie shares his current location with Alice, so the location data type grants access rights also for the "street" data type.

To support the different level of access control, e.g. storing data on the device itself, device-based registry and the sharing of data with remote devices, the policy introduces another relation that describes the level of sharing data. The *sharingLevel*-relation shows on which level data may be shared or stored. Using this relation, the privacy policy can be easily used to enforce the sharing level on data. GAMBAS relies on three pre-defined keywords that describe where data may be stored. The tree keywords are "Remote", which defines that data may be stored by remote devices, "Device", which denotes that the data should only be stored at the device itself and must not be shared with others, and "DetectOnly", which does not allow the data to be stored anywhere. When "Device" or "DetectOnly" are chosen, the *Permission*-relations are ignored, i.e. data may not be shared with anybody, when using this keyword.

An example for the *sharingLevel*-relation is presented in Figure 5.14. Here, location data is shared with remote devices; for the access rights, the *Permission*-relations that are linked with the location data type must be considered. Data about the current travel path may be stored on the device and used for prediction that is executed on the device. This policy triple does not allow sending the current travel path data to remote devices. In this example, audio data might only be used for detection using the DQF, but not be stored anywhere.

In summary, the privacy policy consists of three relation types. All of them can be described using privacy triples:

- The *Permission*-relations that define access rights and obfuscation levels of data types.
- The *consistsOf*-relation that defines hierarchical relationships between data types.
- The *sharingLevel*-relation that defines the sharing level of the data type.

Next, we describe the policy generator, which enables the automatic generation of the policy from social networks or other collaboration tools. Thereafter, we describe the integration of the privacy policy with the user interface to enable the user to modify the policy manually.

```
Location sharingLevel Remote
CurrentTravelPath sharingLevel Device
Audio sharingLevel DetectOnly
```

Figure 5.14   Privacy Policy Sharing Level Example.

5.3.1 Automatic Generation

The privacy preservation policy that is used by the PRF to constrain the access to data gathered by the DQF can be created automatically, by the policy generator. This enables the user to use GAMBAS applications without an extensive (manual) configuration phase, while still having a privacy policy that protects private data. The policy generator is therefore one of the key concepts to enable automation in the privacy preservation framework.

Many users use social networks or collaboration tools like Google Calendar as part of their everyday routine. They post messages to colleagues and friends, share photos and create shared appointments. Often, it is possible to constrain access to messages to a pre-defined user group. This could be a list of friends on Facebook or individual users for a shared event in Google Calendar. Similar, the access to other data in the social networks or collaboration tools can be constrained by the user. Figure 5.15(a) depicts an example. Using the APIs that are provided by the social networks or collaboration tools, these privacy settings can be retrieved automatically. An example using Facebook's graph API is shown in Figure 5.15(b).

A user that is using such a social network or collaboration tool is therefore already creating one or more privacy policies (depending on the number of tools that are used). The privacy policy generator can query these policies using the tools' APIs. Since the policy generator operates on the user's own device(s) and uses the user's accounts to access the tools, a policy which is individual for each user can be generated. This generated policy is then also tailored to the needs of the user, because it is only an import of a user-defined policy into the privacy framework. To support as many social networks or collaboration tools as possible, the policy generator has a modular structure. This structure makes the policy generator extensible with regard to other collaboration tools.

Usually, the user is able to define privacy settings in each social network or collaboration tool individually. Because the user may edit this settings freely and independent from each other, the settings might be inconsistent. The policy generator is then not able to create a consistent policy. If this is the case, the generator can detect and display the conflicting settings and suggest possible solutions to the user. After the conflicts are (manually) resolved, the policy generator creates a consistent policy.

Using the *consistsOf*-relation, the policy generator proposes different generalization strategies to apply the policy to a broad set of data types that can be acquired by the DQF. This enables the generalization of the policy,

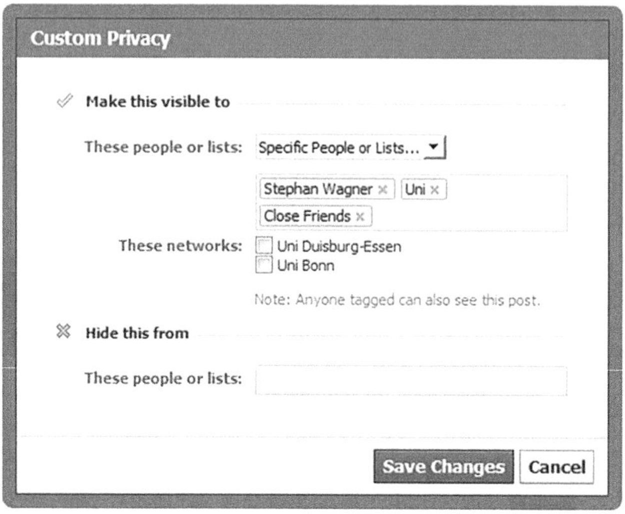

(a)

```
[{"object_id":105482406163077,"id":105482406163077,
"value":"CUSTOM","description":"Uni","allow":"101909386520379",
"deny":null,"owner_id":100001039541728,"networks":null,
"friends":"SOME_FRIENDS"}]
```

(b)

Figure 5.15   Privacy Settings in Facebook. (a) User Interface and (b) Programming Interface (JSON).

which includes new data types that might be related to data types retrieved from the privacy settings in social networks or collaboration tools.

In summary, the policy that is used to allow access to different data types is generated automatically using the policy generator. The generator is designed to pick up policies or privacy settings that are pre-specified by the user in a social network or collaboration tool and to create a policy that is compatible to the GAMBAS policy format. The generator includes tools to resolve conflicting settings and is able to generalize data types. The generated policy can also be fine-tuned by the user using the user interface presented in the next sub-section. Even without the fine-tuning, the generated privacy policy is consistent and tailored to the user's needs, without putting the user's privacy at risk.

5.3.2 Manual Fine-Tuning

The user interface developed as part of the middleware enables the user to fine-tune the privacy policy. In general, the privacy policy is created automatically using the policy generator. The automated creation takes into account the settings of the user in social networks and other collaboration tools, like the Google Calendar. Although this automatically derived policy is therefore created on an individual basis, a user may want to modify the policy. To do this, the privacy preservation framework encompasses methods that allow retrieving the current policy and methods that can modify the existing policy.

Using the user interface, the user can change the policy triples visually, without having to use the policy language. This allows also non-expert users to edit the policy successfully. The user interface displays the data types and then shows the relevant policy relations graphically. The user can modify the data types by clicking on them and, for example, choose users from a list of users for the *Permission*-relations. Editing the other relations is similar. Any change in the graphical user interface results in a change of the policy, i.e. causes the addition, deletion or modification of policy triples. The user might also use the interface to export or import the privacy policy, which enables expert users to modify the RDF representation of the policy directly.

5.4 Privacy Integration

To clarify the mechanisms and protocols of the privacy framework, we describe how they are integrated into data transfer, data acquisition and data processing defined in the previous chapters.

5.4.1 Data Transfer

To support data exchange and possibly the exchange of context information, data will be transferred between different devices. This data – for example, a user asking the servers of a public transit network operator for the route of a bus trip – can breach privacy. In this case, an eavesdropper could get the current and future location of the user. Therefore, the data should be transferred securely. In GAMBAS, the Privacy Preservation Framework (PRF) is responsible for all security and privacy needs and therefore is also responsible for securing the data transfer. For this, all data that is transferred should be encrypted. The reason for this is twofold. Firstly, the data might contain private information that should not be shared with unauthorized users or devices. Secondly, the shared data might be transferred over an insecure communication channel (e.g. the Internet or an insecure WiFi network).

To apply the efficient concept of symmetric encryption (AES) to secure communication, a shared key must be exchanged before any encrypted communication can take place. During the exchange of a cryptographic key, the communication endpoints show that they are eligible to access the data that should be transferred by authorizing themselves. After the authorization process, both endpoints possess a shared cryptographic key that allows them to transfer data securely.

In the GAMBAS PRF, authorization can be performed in two different ways. The first way uses asymmetric cryptography and is based on certificates, similar to the implementation of SSL in the Internet. This allows an ad-hoc identification of devices that belong to a certain domain. If the domain root is trusted, the authorization will be successful. Also, the access rights may depend on the trust in this root. For authentication, the device's certificate is transferred together with a challenge that proves that the device is in possession of the certificate's private key. Together this data forms the device's credentials that are checked at the other endpoint. The alternative of using compute intense asymmetric cryptography is symmetric cryptography. Using symmetric cryptography, a key (256 Bit) can be attached to a connection between two endpoints. The first half (128 Bit) of this shared key allows the identification of the other endpoint. The other half (128 Bit) can either directly used for the secure communication or can be used to exchange a new session key securely. For efficiency reasons, both of these checks (i.e. for asymmetric and symmetric cryptography) are performed transparently by the communication system of the GAMBAS middleware.

The secure data transfer is generally foreseen for every transmission of data. The communication endpoints must first authorize each other at the remote privacy preservation framework, before a key for the secure communication is computed. The authorization that is performed by the privacy-preserving framework incurs some overhead during the data transfer. However, without the authorization, the communication partner is unknown to another device and this contradicts the privacy of the transferred data. Therefore, while the authorization is a crucial mechanism, it is possible to use more lightweight security mechanisms, but this would result in a decrease of the security level.

To enable encrypted data transfer, it is necessary for both communication endpoints to use a cryptographic key. In GAMBAS, we support devices with different capabilities with regard to the available resources (like RAM, CPU and battery power). The encryption is therefore based on a hybrid scheme that allows an efficient and secure encryption. With respect to data processing, we can differentiate between three different cases. The first case is the communication between a user's device (a client) and a server (e.g. the server is asked for a bus route by a user's device). The second case is the communication between devices of two users. An example could be two friends who want to exchange photos with each other. The third case is the communication between two servers. An example could be server of a public transit network operator who communicates with a weather service server to get the current forecast (which might influence the bus planning for the day).

The difference between a user's device and a server in GAMBAS is that the server is able to authenticate itself using a cryptographic certificate. This ensures the identity of the server and allows for server authentication, before the connection is established. In contrast, the user's device does not have a certificate since it is not bound to a certificate domain. Therefore, different methods of authentication must be used, if a user needs to be authenticated and/or identified. Because of these differences, the three cases mainly differ in the authentication phase.

5.4.1.1 Client and Server Communication

When a device (client) contacts a service that is provided by a server, the connection will be established as shown in Figure 5.16. The server will use its certificate to authenticate itself against the client device. The client application validates the certificate of the server against either a pre-deployed service certificate or a pre-deployed certificate root. The root certificate can be used for companies that support different kinds of services and eases the deployment without lowering the provided security. This is a one-sided authentication, i.e. the server authenticates itself against the client, but the client is not authenticated. If it is necessary to authenticate the client, the server may add any authentication scheme after the secure connection is established. A username/password authentication could, for example, use the secure remote password protocol (SRP 6) to authenticate the clients.

Using the (unencrypted) authentication messages, an elliptic curve Diffie–Hellman (ECDH) key exchange is performed. As a result, both sides will be able to compute a secret key that cannot be computed by eavesdroppers. The signature of the message that is created by the server also prevents man-in-the-middle attacks. After the authentication handshake and the key exchange, the interaction is encrypted by using the exchanged secret key.

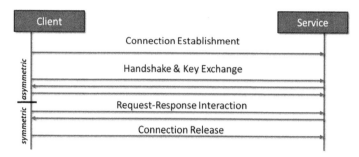

Figure 5.16 Client and Server Communication.

5.4.1.2 Device-to-Device Communication

When two devices establish a connection, usually they cannot authenticate each other. The authentication cannot rely on certificates, since the devices do not know each other and normally do not possess a certificate. The users might identify their devices manually, by device id, and may then create a manual key on both of them. But GAMBAS also allows two other easier methods to establish a key for each device.

The first method uses PIKE to exchange a key using an online social network such as Facebook that is used by both users. This key is exchanged before the interaction takes place and allows two or more devices to interact securely with each other. The user identification is extracted from the relationship in the online social network and can then be used at the time later on. This allows for a completely automatic key exchange that does not need any user interaction. The only necessary step a user has to take is to connect GAMBAS with the social network, e.g. through Facebook Connect as shown in Figure 5.17(a).

The second method uses the NFC technology. Nowadays, many smartphones are equipped with an NFC reader system that may also be used for short-range one-way communication. The GAMBAS middleware implementation for Android integrates with NFC to exchange a key. For this, the middleware encompasses a service that is used to redirect the communication back to the device that initiated the one-way NFC communication. This backward channel is necessary to transfer the device id of the device that received the NFC message. Using this service, we can establish a key just by holding two devices together. For this, a user simply needs to press a button in the user interface shown in Figure 5.17(b) and then bring two phones in physical proximity to each other.

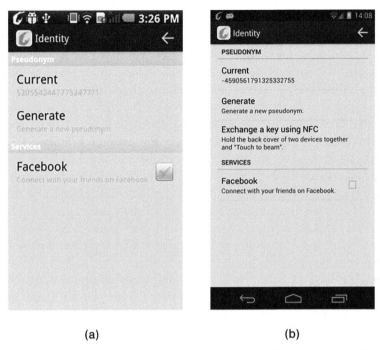

(a)   (b)

Figure 5.17   Device to Device Authentication. (a) PIKE-based Key Exchange via Facebook and (b) Manual Key Exchange via NFC.

After the successful key exchange, the communication between the two devices can be encrypted. Additionally, the devices can identify themselves using the already established keys.

5.4.1.3 Server-to-Server Communication

The server to server communication in GAMBAS is similar to other communication on the Internet. Each server authenticates to the other server using its own, pre-deployed certificates. Similar to the client–server communication, the certificates can be verified by using the server certificates or by using a common root certificate. Again, the authentication includes an ECDH key exchange, which results in a shared key for this connection. Then, the communication between the two servers will be encrypted using this key.

5.4.2 Data Acquisition

The Adaptive Data Acquisition Framework (DQF) enables the collection of data using various sensors built into the user's mobile device. The collected data can then be used personally (i.e. by the device, in the case of personal data acquisition) or collaboratively (i.e. by a remote service, in the case of collaborative data acquisition) to optimize services based on the users' behavior. Clearly, the data acquired by means of sensors built into the device of a user may raise privacy concerns. Furthermore, the preferences with respect to privacy may vary drastically from user to user. In order to empower users to exercise control over which data can be collected, the access to the data acquisition framework is guarded by the Privacy Preservation Framework (PRF). Thereby, all accesses made to the data acquisition framework are checked against the user's privacy preferences with respect to data collection. This allows the user to limit the data types that can be collected at all. In extreme cases, a user may limit the collection of all data through the GAMBAS middleware. In less extreme cases, the user may limit the collection of a particular type of context information, such as location-related information or audio information.

The PRF-DQF interface enables the data acquisition framework to check whether the user has given consent to the acquisition of a particular type of contextual information. To do this, the DQF performs calls to the PRF in order to verify that the data types that shall be captured are permissible under the user's current preferences. Furthermore, since the user's preferences may change at any point in time, it is necessary that the PRF provides functionality to signal a change to the DQF whenever the user's preferences with respect to a particular data type change.

The PRF therefore has two different duties. First, it checks the data type that is about to be captured against the preferences of the user and returns a Boolean to indicate whether the user permits the acquisition of the specified data type. If the access is denied, the acquisition is aborted. If access is granted, the acquisition task can be started. Additionally, a user could modify his privacy settings. Therefore, the PRF needs to signal a change to the preferences with respect to a particular data type such that the DQF can check all currently executed data acquisition tasks against the updated set of preferences. If a data acquisition task is no longer permitted by the user, it must be aborted by the DQF.

In order to guarantee that all data acquisition tasks continuously conform to the user's preferences, the GAMBAS middleware implements the continuous and gapless usage of this interface for all calls to the DQF. This means that all tasks that are started within the DQF need to pass through the check method of the PRF with the associated data types. In addition, as long as the DQF is executing any tasks, it needs to react to changes indicated by the signal method. If a signaled change affects a data type that is currently acquired, the check for the associated (set of) task(s) needs to be reevaluated, possibly aborting any conflicting tasks.

As every export of user's context information is filtered based on the privacy policy of the user, for every request of data by the service provider, the DQF checks it with the privacy framework. If the PRF allows the data to be sent, only then the users' context information is exported. The PRF and DQF communicate this information check through control interfaces provided by the PRF. Specifically, PRF provides different methods that allow the DQF to check if a certain data type is allowed to be detected. The DQF must call these methods before creating a recognition stack for detecting any kind of data or context. Based on the results from these methods, the DQF detects the context data and subsequently sends it to the service providers.

In order to allow acquisition and subsequent export of user's context information, the GAMBAS middleware ensures that the context recognition applications can gather and export only the allowed context features. In order to achieve this, the data DQF checks for permissions with the privacy-preserving framework whenever a new application is started.

When the data acquisition framework starts to acquire data, it analyzes the feature requirements of the application and then checks with the PRF whether the desired features are allowed to be gathered. The PRF will decide, based on the privacy policy that is set by the user and will inform the acquisition framework whether the requested features are allowed to be gathered or not. If the requested features are allowed, then the DQF starts gathering context information.

When the PRF refreshes the privacy policy (either through a user that edits the policy or through an update issued by the Privacy Policy Generator), the GAMBAS core service indicates this change to the DQF, which again checks the permissions with the privacy framework. If an already running application does not adhere to the new privacy policy, then the application is shut down immediately.

The list of privacy features that a user can edit in the privacy policy includes features related to the acquisition of sensing data such as audio sensing, location sensing, motion sensing, ambient sensing and features related to the communication such as enabling of remote gateway communication, enabling of Wi-Fi and Bluetooth as communication technologies, etc.

5.4.3 Data Processing

Dynamic and distributed data processing is an essential part of the GAMBAS middleware. Data processing in GAMBAS is performed by the Query Processors (xQP), which provides GAMBAS applications with the necessary data. Often, the processor executes remote queries. These queries are executed at remote devices and may try to access private data. The GAMBAS Privacy Preservation Framework therefore has to check the access to the requested data types and allow/deny access based on the policy of the remote user.

During the query execution, the query processor identifies the sources needed to answer the query and then sends a request to the registry. The registry resolves the sources and sends back to the processor the list of endpoints (remote storages) that contain needed data. For shared data, however, before the query processor can access the data on the remote source, a privacy control is performed to check if the query initiator has the rights to access the data. A view of the data matching the privacy rules in place is created and shared with the query processor. The query processor forwards the identity and data requirements to the privacy framework, which in turn checks with the privacy framework of the remote device hosting the shared data. A view of the data is created based on the access control. The view can reflect the original data, or it can modify the original data according to the privacy in place. For example, it can aggregate or hide parts of the original data, like changing GPS coordinates to the name of the city or country.

If a one-time query is issued, the access is granted based on the privileges of the user that is trying to access the data. The query can then be directly executed and will be transferred over a secure connection. If a continuous query is issued, a secure access token is generated and sent to the query processor. If a remote endpoint is trying to access the shared data, the secure access token will allow transferring the shared data securely over the chosen communication channel.

The interface between the xQP and the PRF checks whether the query initiator is allowed to access the data. Additionally, if the xQP is executing a remote query, the communication must be properly secured. The user and data access credentials are sent over a secure data connection between the two endpoints. Since the middleware manages the secure communication transparently, the interface does not include a method that enables the exchange of security tokens or start the encryption. Instead, this is done through the authentication and key exchange plug-ins that the PRF integrated into the GAMBAS middleware.

The access to data by the query processor must be checked through an interface at the PRF. The interface consists of one function that checks if the query initiator (i.e. the user requesting the data) is allowed to access the data. The data types that are being requested also need to be specified. The PRF queries the privacy policy of the device using the specified input and decides whether the query is allowed or not. Each request is handled by the privacy framework of each semantic data storage; therefore, this function is performed locally.

The PRF therefore has a local PrivacyManager that implements and interface that can be used by the xQP to check with the PRF if executing a received query is allowed according to the currently active privacy policies. To do so, the query processor hands the PRF (1) a set of classes in the GAMBAS ontology that specify what data types the query will access and (2) the origin of the query, e.g. if it was a local query or a query from a remote user. The PRF then returns whether this query is allowed or not. The PRF needs to be contacted for every query execution, when shared data is involved. The query processor must first interact with the privacy framework, which is responsible for allowing or denying data access, for data encryption/decryption and for device authentication.

Of course, the privacy-preserving framework incurs some overhead in the query processing, specifically an additional method call, device authentication and data encryption. However, the PRF is crucial to maintain the privacy of the users' data. To minimize the performance impact, the PRF uses lightweight privacy rules and lightweight encryption mechanisms (e.g. symmetric encryption using AES) to allow a secure and privacy preserving execution of queries by the xQP. More lightweight encryption mechanisms could be applied, but this would result in a decrease of the privacy and security level without a high speed-up compared to the used security mechanisms, if measured on current smartphones.

6

Applications

This chapter describes the applications that have been built using the GAMBAS middleware. To do this, the chapter briefly outlines the integration of the system components described in Chapter 3, Chapter 4 and Chapter 5. Based on this description, it introduces the application development support provided by GAMBAS for different execution environments. To clarify this, we present a number of simple but full-featured applications that leverage the different components of the middleware. Based on this, we then describe the two large-scale applications that have been built with the middleware. These applications focus on realizing significant parts of the mobility scenario and the environmental scenario introduced in Chapter 1 that motivated the work on the GAMBAS middleware.

6.1 Application Development Support

In the following, we describe how the GAMBAS middleware is used during application development. To do this, we first briefly review how the different middleware components described in the previous chapters are integrated into a single system. Thereafter, we discuss how different execution environments are supported through the GAMBAS SDK and middleware runtime. Finally, we present a number of simple applications that have been built with the SDK to demonstrate the different features offered by GAMBAS.

As shown in Figure 6.1, the integrated GAMBAS system consists of (1) a number of networked devices executing the GAMBAS middleware and (2) the GAMBAS Dynamic Data Registry. Each device may execute one or more GAMBAS applications (or simply apps) using the GAMBAS middleware. An example for such an app is an Android application executed by an end user on his smart phone. Another example would be server software

161

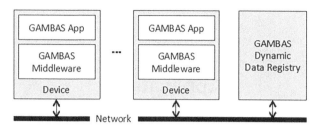

Figure 6.1   Integrated System.

executed by a service provider on a dedicated server connected to the Internet. As described in Chapter 4, the GAMBAS Dynamic Data Registry is a generic service that provides devices with the ability to discover data sources. The functionality of this registry is comparable to the Domain Name System (DNS), which provides name resolution on the Internet. Although GAMBAS assumes that this functionality is provided publicly, GAMBAS allows developers to run their own registry during development and testing. Since the functionality of the registry has been described in detail in Chapter 4, in the following, we focus on the remaining functionalities.

6.1.1 Overview

Figure 6.2 gives an abstract overview of the middleware structure. The integration is realized by: (1) a set of interfaces, support libraries and tools called the *Software Development Kit* (SDK) and (2) the *GAMBAS CoreService* which provides the accompanying runtime environment. The Software Development Kit (SDK) in turn consists of two parts. The *Service Programming Interface* (SPI) is used to develop GAMBAS functionality and integrate it into the middleware. The *Application Programming Interface* (API) is used to develop GAMBAS apps. The CoreService sets up the GAMBAS middleware and manages the life cycle of GAMBAS system components. Each system component encapsulates the implementation of one of the core GAMBAS parts described in Chapter 3, Chapter 4 and Chapter 5, e.g. the Semantic Data Storage (SDS) or the Data Acquisition Framework (DQF).

In addition, the CoreService integrates a special communication system component that encapsulates an extended version of the BASE communication middleware discussed in Chapter 5. This enhances GAMBAS with communication support to interact with remote GAMBAS devices. Furthermore, the CoreService realizes the SPI by linking each system component to all other components that they use during their own execution via interfaces from

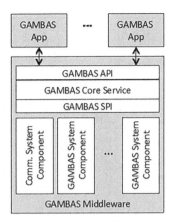

Figure 6.2   Abstract Middleware Structure.

the SPI. This effectively provides a tight and efficient integration between the components without inducing dependencies to their actual implementation. Finally, the CoreService implements the GAMBAS API towards GAMBAS applications in both Android and J2SE environments. To do this, it receives calls, forwards them to the right system component and delivers results back to the original caller.

Due to the intrinsic differenced between Android and J2SE execution environments, the abstract structure shown in Figure 6.2 has two distinct concrete implementations. In the following, we briefly describe their differences and similarities.

6.1.2 J2SE Support

GAMBAS for J2SE specializes and implements the generic middleware architecture described before for server systems running J2SE. This allows service providers to integrate their services into the GAMBAS platform. Figure 6.3 shows the resulting system architecture. Since the J2SE version of GAMBAS is primarily intended for the development of server systems, it does not include support for user interfaces. Clearly, service providers will, in many cases, add their own user interface, e.g. based on web technologies. This, however, is outside of the scope of GAMBAS and thus not explicitly supported. All other GAMBAS system components mentioned previously are integrated, namely communication, data acquisition (the DQF), data storage (the SDS) and querying (the xQP), as well as security and privacy (the PRF).

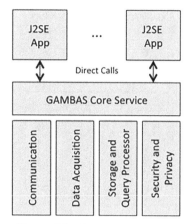

Figure 6.3   GAMBAS for J2SE.

As described before, the CoreService realizes the SDK and manages the life cycle of the whole GAMBAS system and all its components on a local device. GAMBAS for J2SE is implemented as a library that is linked to an application using it. To start using the GAMBAS system, an application has to first import and instantiate the CoreService. The CoreService can be configured by passing it an instance of *CoreSetting*. This allows the application to specify, e.g. the address of the GAMBAS data registry and communication gateway as well as a pseudonym that should be used to address the system. Settings can be changed dynamically and the CoreService will perform any necessary updates automatically, e.g. when a pseudonym should be changed. When the CoreService is instantiated (and thus started) it instantiates, configures and starts in turn all necessary GAMBAS system components.

To decouple life cycle management of components from their actual implementation, each component is encapsulated by a specific subclass of *AbstractSystem*, providing, e.g. methods for startup and shutdown. As an example, the SDS is integrated by sublassing AbstractSystem with a new class *DataStorageSystem*, which implements all life cycle management functions independently of the actual SDS implementation. This way, the SDS is independent of the CoreService and can, e.g. be reused in other contexts without other GAMBAS components. The CoreService also passes each component references to all other components it may require. As an example, the query processor uses the data storage, the communication system and the privacy manager. The CoreService enables this by passing references to these three components to the query processor. At runtime, the query processor calls these components directly, without using the CoreService anymore.

To use functionality of the GAMBAS middleware, e.g. to store data, applications can call a number of methods on the CoreService. The Core-Service in turn forwards this request to the corresponding local system component, retrieves results from it and forwards them to the calling application. This design was chosen over directly exposing system components to applications because it allows the CoreService to impose further checks on the correctness, security and consistency of these calls, if needed. As an example, the CoreService might deny a new request if it has already started shutting down the system. In addition to this, this approach also provides access transparency for system components, i.e. it allows us to decouple the system components from the way they are called. If an application wants to call a remote system component (e.g. to store data in a remote SDS), it can do so by calling a local method on the CoreService. The CoreService will forward this request to the communication system (essentially acting as a communication broker), which will send it to the remote system. There, the incoming request will be forwarded by the communication system to the CoreService, which in turn forwards it to the corresponding system component. Finally, if an application wants to stop the GAMBAS middleware, it again calls the CoreService, which notifies all components and shuts down the system correctly.

As a result, the CoreService is the central component of this architecture. It is responsible for receiving and answering all local and remote calls from applications, mediates all dependencies between system components and fully manages their life cycles. This encapsulates nearly all integration activities into it, reduces the complexity of implementing the actual system components and allows them to focus on their core functionality.

6.1.3 Android Support

In addition to J2SE devices, the GAMBAS middleware also directly supports application development on Android devices. From a high-level perspective, the Android integration is similar to the J2SE version, but when looking at the details, it has two main differences: first, it separates the core middleware from applications using it, reflecting the distinct Android runtime model and reducing the overall resource need of the system. Second, it includes additional support for user interactions with the intentional user interface (IUI). The resulting architecture can be seen in Figure 6.4.

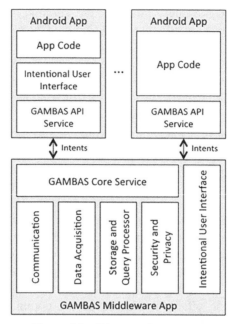

Figure 6.4   GAMBAS for Android.

6.1.3.1 GAMBAS Middleware App

On Android, the main functions of the GAMBAS middleware are realized as a stand-alone Android app instead of a linkable library. This app is independent of any third-party Android applications using it. On Android, the life cycle of an app is controlled by the OS. It may at any time pause or stop/destroy any app, if it requires more resources. If the GAMBAS middleware would be linked to an app using it, the OS could decide to stop it, if the app is not used by the user right now. By separating the middleware into its own app, we are separating its life cycle management from that of all apps that use it. In addition, this design allows us to efficiently share a single instance of the middleware between all third-party apps, reducing the needed resources and thus allowing the OS to keep all apps active in memory for a longer time.

An alternative approach would be to model the middleware as an Android service. However, allowing the middleware to have its own user interface – independently of any other app – allows us to integrate all configuration activities that a user wants to perform for the whole system in one place. In addition, it also allows the user to start and stop the middleware explicitly, since its execution will reduce battery lifetime. To remind the user that GAMBAS is running, we display a corresponding icon in the Android status bar. By clicking this icon, the user can display the middleware user interface and control its behavior, e.g. reconfigure or stop it.

The separation of the middleware into a distinct app also influences how third-party apps can access its functionality. Direct calls are no longer possible since the apps are running in separate processes. Therefore, we use Android intents to interact with the middleware. Intents are small events or messages that a process can publish and that can be received by other processes. To use the GAMBAS middleware, an application can publish a number of intents that are received by the middleware. The CoreService includes support for this. It translates the intents into direct calls and forwards them to the corresponding GAMBAS system components. Once a result is available, the CoreService translates it back into an intent and publishes it, allowing the original app to receive it.

Clearly, this is more complicated for app developers than directly calling a method on a Java object. To reduce the complexity of the interface, we provide a GAMBAS API service. This service is realized as a Java class that can be subclassed by an application developer. It already includes all necessary functionalities to translate calls to the middleware into intents and vice versa as well as additional support for handling life cycle and error. Thus, by using this API service, the app developer can access the middleware without knowing about the specifics of Android interprocess communication.

6.1.3.2 GAMBAS User Interface

As described above, the GAMBAS middleware for Android devices contains the Intent Aware User Interface (IUI). The IUI is separated into two parts: an interface to control the behavior of the GAMBAS middleware itself and support for the development of user interfaces of third-party apps.

The GAMBAS middleware user interface enables the user to configure a multitude of aspects (Figure 6.5(b)) such as the middleware life cycle (Figure 6.5(a)), the used data discovery registry and communication gateway (Figure 6.5(f)), the user's pseudonym, known friends, their keys (Figure 6.5(d)) and privacy policies. This allows users to inspect and adapt the current system state in one integrated place and makes it much easier for them to understand what data is currently made available to whom.

In addition, the middleware user interface allows to manage all third-party GAMBAS apps (Figure 6.5(c)) in an integrated view. This view allows to

Figure 6.5   User Interface. (a) Start, (b) Settings, (c) Apps, (d) Privacy, (e) Features, (f) Development.

install new apps from the Google Market, to start them and to remove them once they are no longer needed. Finally, in order to give full control over sensing to the users, the user interface also enables users to disable the different data collection components offered by the data acquisition framework.

6.1.4 Application Examples

To test and showcase the GAMBAS SDKs, we have developed several applications that demonstrate the use of the different features of the GAMBAS middleware. These application have been made available to developers and they have also been published on the Android market. In the following, we briefly describe three of these applications. We first describe the application functionality and then map it to the middleware functionality.

6.1.4.1 GAMBAS Voiceprint Launcher

The GAMBAS Voiceprint Launcher is an Android application developed on top of the GAMBAS middleware. The application uses the voiceprint technology developed as part of the data acquisition framework (c.f. Chapter 3). The Voiceprint Launcher enables a user to launch an application by issuing a voice command. To enable the launching of applications via a voice command, the user first needs to train the launcher by creating recordings of the commands that shall start different applications (see Figure 6.6).

To do this, the user can add an application from the list of applications installed on the device. Then, the user can select the application and press the train button (i.e. the button with the white headset) to start the training. Alternatively, the user can press the delete button (i.e. the button with the white trash can) to delete the application and all training data. Once the

Figure 6.6 Voiceprint Luncher Training.

user has pressed the train button, a dialog appears that prompts the user to say the application name loud. Once the user completed this, the application computes a voiceprint and stores it locally.

As soon as the user has trained one or more applications, he can start the application by pressing the start button (i.e. the white microphone). This will open up a dialog that prompts him to say the application name out loud. Once he has done that, the application will compute a voiceprint and match it against all stored voiceprints. The closest match will be selected and the associated application will be started. Alternatively, the user can also add the voiceprint launcher widget to the home screen of the device. This allows the user to directly access the launcher (see Figure 6.7).

From a technical perspective, the GAMBAS Voiceprint Launcher demonstrates a substantial part of the middleware. However, it is noteworthy that it solely executes locally on the phone of a user and thus, it does not require any remote connectivity or services. Consequently, it does not cover any communication-related aspects and it also does not cover the J2SE integration. As depicted in Figure 6.8, the GAMBAS Voiceprint Launcher extensively uses the GAMBAS middleware on Android through the Android SDK that connects it with the core service of the GAMBAS Middleware App.

Of the functionality provided by the core service, the GAMBAS Voiceprint Launcher uses four out of five building blocks as follows:

Figure 6.7 Voiceprint Launcher Usage.

Figure 6.8 Voiceprint Launcher Coverage.

6.1.4.1.1 *Data Acquisition*

In order to capture audio data and to compute and classify voiceprints, the application uses the audio components of the context recognition framework. In particular, the application uses an AudioSensor component to capture audio, a Windowing and FFT component to perform preprocessing, a trigger component for silence detection, a voiceprint generator and matcher component for voiceprint computations and classification as well as an intent broadcaster component to signal the successful acquisition of a voiceprint as well as to signal the classification result. Figure 6.9 depicts the configurations of the component system.

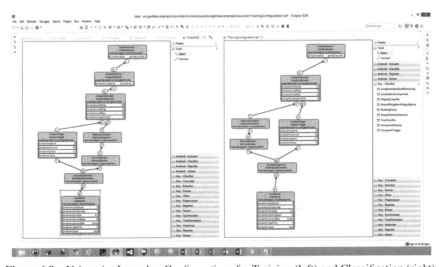

Figure 6.9 Voiceprint Launcher Configurations for Training (left) and Classification (right).

6.1.4.1.2 *Data Processing*

To store the voiceprints as well as the set of configured applications, the GAMBAS Voiceprint Launcher uses the semantic data storage as well as the SPARQL-based query processor. To store the voiceprints, they are serialized as strings such that they can be stored as RDF triples. To retrieve the set of configured applications and the associated serialized voiceprints, the GAMBAS Voiceprint Launcher uses the SDK to issue SPARQL queries against the data storage that are executed with the middleware's built-in local one-time query processor.

6.1.4.1.3 *Privacy Preservation*

Although the Voiceprint Launcher is executed locally on the device, it still integrates with some of the privacy features of the GAMBAS middleware. In particular, as depicted above, the GAMBAS Voiceprint Launcher's access to the device's soundcard and audio capabilities are controlled through the GAMBAS middleware. Thus, a user can prevent the application from recording audio by simply deactivating the associated middleware feature. Consequently, the requests to capture audio by means of the configurations depicted previously will be blocked by the middleware. The associated blocking will then be signaled back to the application via the Android SDK such that it can react to it in an adequate way, for example, by showing a dialog that tells the user that the application requires audio capabilities to function properly. Intuitively, for more complex applications, it may also be possible to provide different modes of operation, e.g., a mode that uses audio and a mode that does not. However, for the GAMBAS Voiceprint Launcher, the ability to record audio is essential. Consequently, it is not feasible to provide such a mode.

6.1.4.1.4 *Intentional User Interface*

Similar to security and privacy, the GAMBAS Voiceprint Launcher also integrates with the intentional user interface through the SDK. In order to provide the user with a clean view on the applications that are installed as well as the features that are requested by them, the GAMBAS Middleware app uses intent-based interaction to populate the list of installed GAMBAS-enabled applications. This enables the user to quickly list all GAMBAS applications and to start an application from the GAMBAS Middleware App's user interface.

6.1.4.2 GAMBAS Linked Weather

While the GAMBAS Voiceprint Launcher is focused on the Android SDK, the GAMBAS Linked Weather application focuses primarily on data management, remote communication and the J2SE SDK. The application uses the legacy data wrapper to integrate with a third-party data source, namely the weather web service provided by Wetter.com. To do this, a J2SE-based service periodically retrieves the weather information for the largest German cities and stores it in a semantic data storage that is equipped with remote communication and distributed query processing functionality such that the data becomes accessible to other devices.

To demonstrate the J2SE application as well as the interaction between J2SE-based and Android devices, we have developed a Linked Weather app. The functionality provided by the application is depicted in Figure 6.10. The application enables a user to add an arbitrary number of cities to his device. Once the cities are added, the user can press a sync button to retrieve the latest weather information. Internally, tapping the sync button will issue a series of remote SPARQL queries against the RDF data stored in the semantic data storage on the J2SE device, which will synchronize the local data storage of the Android device with the remote data storage of the server. In order to reduce the amount of data that must be transferred, however, only the cities selected by the user are actually synchronized. When the synchronization is completed, the user can tap any city to view the current forecasts. This will issue a series of local queries against the storage, to retrieve the forecasts for a city. At this point, there is no more need for remote interaction as the device already has the associated data.

As depicted in Figure 6.11, the GAMBAS Linked Weather application uses the GAMBAS middleware on Android through the Android SDK and on J2SE through the J2SE SDK. From the functionality provided by the core services, the GAMBAS Linked Weather uses four building blocks as follows:

6.1.4.2.1 *Secure Communication*

In order to interact with each other, both the Android and the J2SE parts of the application use the communication services provided by the middleware. Thereby, the Android part of the application contacts an application-specific service provided by the J2SE part of the application. To determine the communication endpoint that provides the application-specific service, the Android app interacts with the dynamic data discovery registry.

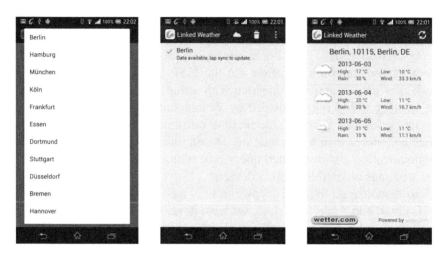

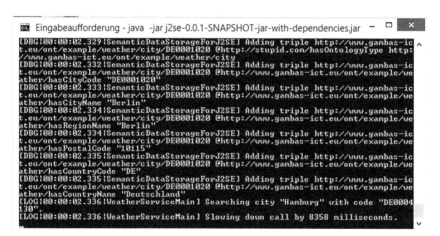

Figure 6.10 Linked Weather Android App and J2SE Service.

6.1.4.2.2 *Data Processing*

In order to store the weather information, both the Android and the J2SE parts of the application use the storage facilities provided by the local semantic data storage. In addition, the Android part of the application executes (remote) queries on the semantic data storage of the J2SE part of the application in order to synchronize the local weather information with the most current version of the weather information provided by the J2SE application.

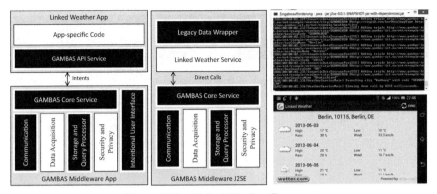

Figure 6.11 Linked Weather Coverage.

6.1.4.2.3 *Intentional User Interface*

Similar to the GAMBAS Voiceprint Launcher, the GAMBAS Linked Weather application also integrates with the intentional user interface through the SDK. The integration closely follows the explanation given previously in the sense that the application is shown in the associated list with the associated permissions.

6.1.4.2.4 *Legacy Data Wrapper*

In order to integrate with Wetter.com, the actual provider of the weather information made accessible through the J2SE service, the J2SE specific part of the application uses a legacy data wrapper that translates the custom data model used by Wetter.com to linked open data that is then stored in the semantic data storage and made available through the query processor to mobile devices. To gather data, the J2SE service periodically pull the latest data from the web service. However, in order to avoid exceeding the free quota provided by Wetter.com, the pull frequency is set to one day.

6.1.4.3 GAMBAS Locator

To demonstrate the location prediction algorithms developed as part of the data acquisition framework as well as the privacy-preserving data-sharing among devices, we have developed the GAMBAS Locator application depicted in Figure 6.12. Similar to the GAMBAS Voiceprint Launcher, the GAMBAS Locator only uses the Android version of the GAMBAS middleware. From an end-user perspective, GAMBAS Locator enables users to continuously track their location. They can track visits to locations that are relevant for them and they can share their current location with their friends

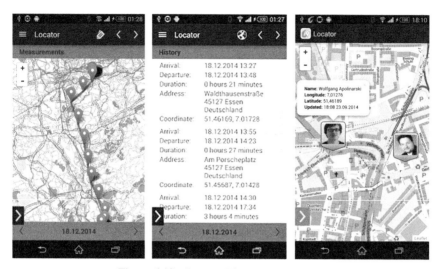

Figure 6.12   Locator History and Sharing.

in a peer-to-peer fashion through the GAMBAS middleware. In addition, the application computes and visualizes predictions for the next user location based on the location history captured by the application. Since the history and predictions are stored in the local SDS of the user's device, they can be used by other applications easily, i.e. by simply querying the local SDS.

To enable the sharing of location information with other users, the GAMBAS Locator uses the secure communication and data sharing mechanisms described in Chapter 5. To perform the necessary key-exchange for user authentication, the GAMBAS Locator can leverage the keys provided by the GAMBAS Middleware App. This means that if a user is using some social network like Facebook, for example, the user simply needs to connect the GAMBAS Middleware App with his Facebook account. Once this is done, the GAMBAS middleware will automatically exchange keys with all of his friends who are also using GAMBAS. If the user does not use social networking sites, he can alternatively use NFC to manually exchange a key. From an application programmer's perspective, using this functionality does not require a single line of code, since the middleware takes care of implementing it. Similarly, in order to share the location information with another user, the GAMBAS Locator does not require any backend service. Instead, due to the distributed processing capabilities of the middleware, the devices can exchange this information directly without a trusted third party. From an application developer's perspective, this eliminates the need and cost for developing and running a service infrastructure. Thus, using GAMBAS, the developer can focus soley on implementing the user-facing functions.

6.1.4.3.1 *Secure Communication*

In order to interact with each other, the Android applications of different users are relying on the secure communication services provided by the middleware. Thereby, the authentication and encryption is done transparently for the application. In addition, it is noteworthy to point out that the sharing is not mediated through a service, which is the common realization of most location sharing apps that are available today. The keys that are required to ensure a proper end-to-end authentication of different users are provided by the mechanisms of the privacy preservation framework.

6.1.4.3.2 *Data Acquisition*

To capture the user's location, the GAMBAS Locator application makes use of the data acquisition framework. To do this, it sets up an component configuration with an associated state machine in the activation system. Together, the component and the activation system perform a periodic but energy-efficient localization of the user's device. To do this, the localization stack integrates the motion sensors, the GPS receiver and the network hardware. This ensures that the energy-hungry GPS receiver is only used when the user's location cannot be established through the motion sensors or through Wi-Fi scans. In addition, the GAMAS Locator also uses the data acquisition framework to perform the predictions on the user's next location. The predictions are triggered periodically whenever a new location is detected. Towards this end, the prediction components are accessing the user's location history that is stored in the semantic data storage of the user's device.

6.1.4.3.3 *Data Processing*

In order to store the location information including the user's location history and the predicted next location, the GAMBAS Locator leverages the Semantic Data Storage of the device. When a prediction must be computed, the prediction components query the local data storage for an (aggregated) view on the user's history. The data model presented in Chapter 4 specifically addresses this issue by supporting aggregation.

6.1.4.3.4 *Privacy Preservation*

To enable access control on the data stored in the SDS, the issuer of remote queries must be authenticated. To do this, the privacy preservation framework

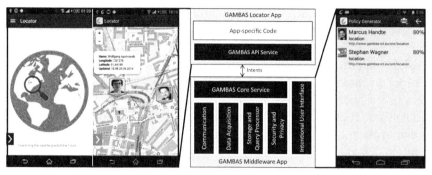

Figure 6.13   Locator Coverage.

presented in Chapter 5 defines two key exchange mechanisms that are either automatic (i.e. when using Piggybacked Key Exchange on top of an online service) or easy to use (e.g. when using a physical gesture to exchange a key between two nearby devices through NFC). In addition to authenticity, however, it is also necessary to define who should be able to access the data. For this, the privacy preservation framework provides an automatic policy generation tool that provides a pre-configured privacy policy based on the user's sharing behavior. Based on this, the user can get recommendations (c.f. Figure 6.13) for suitable policies that can be customized later on. Due to these two mechanisms, the GAMBAS Locator application does not need to handle the intrinsics of secure sharing. Instead, it simply relies on the GAMBAS middleware which automatically provides the necessary mechanisms to enforce the level of privacy desired by the user.

6.1.4.3.5 *Intentional User Interface*

Similar to the other applications, the GAMBAS Locator also integrates with the intentional user interface through the SDK. The integration closely follows the previous explanations. However, due to this integration, the application developer does not need to provide user interfaces to configure the sharing of location information. Instead, the application can simply rely on the definitions managed through the user interface of the GAMBAS Middleware app.

6.2 Application Architecture

As basis for the description of the application components in the next section, we provide an instantiation of the high-level architecture detailed in Chapter 2 for the mobility and environmental scenario outlined in Chapter 1. For each of the scenarios, we describe deployment that maps the abstract software components detailed in the component view to concrete systems. Furthermore, we outline the interactions that will take place at runtime.

6.2.1 Mobility Scenario

As described in Chapter 1, one of the motivating application scenarios behind the GAMBAS middleware is support for mobility applications in a smart city. To demonstrate the middleware capabilities, we developed a so-called Public Transport Exploitation System (PTES) and a GAMBAS mobile application to take into account the information retrieved directly from the user and to offer citizens customized services – not exclusively related to mobility though – in order to enhance their trip experience. Overall, the scenario encompasses personal mobile Internet-connected objects such as the smart phones of citizens, buses that are equipped with embedded systems, existing external services and a number of novel GAMBAS services. Figure 6.14 shows both their deployment and interaction.

6.2.1.1 System Deployment

As depicted in Figure 6.14, the mobility scenario contains a number of computer systems that run various parts of the GAMBAS middleware as well

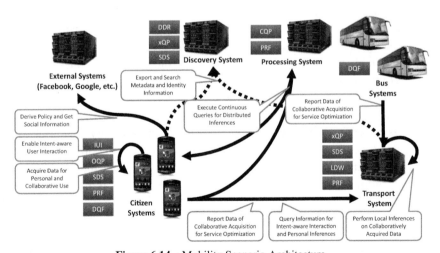

Figure 6.14   Mobility Scenario Architecture.

as application-specific code that realizes the selected use cases. The computer systems are:

- **Citizen Systems:** In order to access services, citizens make use of their personal mobile devices like smartphones and tablets, which are running a mobile application. The application consists of the intent-aware user interface as well as background services that automatically acquire data and either forward or store it. Furthermore, it makes parts of the stored data accessible to other devices. Typically, these systems can be considered as Constrained Computer Systems (CCS). Consequently, the background operations must optimize their resource usage, especially in terms of energy. The application makes use of the data acquisition framework, the semantic data storage and – to protect the user's privacy – the privacy preservation framework. In order to make data available to other devices and to support local inferences at the application level, the device is equipped with a one-time query processor. Finally, in order to enable intent-aware user interaction, the application makes use of the intent-aware user interfaces.
- **Transport System:** In order to provide route information and to aggregate capacity-related information, we introduce a transport system that is available on the Internet. Since the data about bus routes and schedules is already available in a legacy system, the transport system uses a legacy data wrapper in order to tap into this information source. Furthermore, in order to store information coming from the citizen systems as well as from public buses, the system is equipped with semantic data storage. To support local as well as distributed inference on the data and to make it available to third parties, the system is equipped with a one-time and a continuous query processor. Finally, in order to protect the raw data and to restrict the sharing of data, the system is equipped with a privacy framework component that limits the sharing accordingly.
- **Bus Systems:** Besides the citizen systems, the mobility scenario also relies on embedded systems deployed in public buses in order to collect data. Consequently, the buses are equipped with an application that determines the relevant context and forwards it to the transport system, which then stores and aggregates the data. In order to do this, the embedded system running in the bus makes use of the data acquisition framework in order to acquire and report the data.
- **Discovery System:** To enable transparent access to data coming from different data sources, it is necessary to make the possible data sources

discoverable. Performing this task is the primary function of the discovery system. In order to do that, it runs a data discovery registry which uses the semantic data storage component and a one-time as well as a continuous query processor component in order to store metadata and identity information of data sources. In contrast to other systems in the architecture, this system is application-independent.

- **Processing System:** To enable the citizen systems to run continuous queries against each other's devices, the architecture encompasses a second generic type of system. This processing system is equipped with a privacy framework and a continuous query processor.
- **External Systems:** To reduce the configuration effort for the privacy mechanisms, the privacy preservation framework taps into the information available in other external systems. For this, the privacy framework provides a number of adapters that can access the user-specific information in these external systems. Since these systems are maintained by third parties, no additional GAMBAS software is installed on them. Consequently, the adapters of the privacy framework are responsible for performing the necessary data conversion.

6.2.1.2 System Interaction
In order to implement the mobility scenario, the systems and their associated components have to interact with each other locally (within a single system) and some of them have to interact remotely. This interaction follows the abstract interaction patterns described as part of the dynamic perspective in the high-level architecture presented in Chapter 2.

In order to enable distributed query processing, all semantic data storage components export metadata and/or identity information to the discovery system. The query processors and the privacy framework use this information transparently to determine and contact the appropriate data sources and to create the necessary views, respectively.

For the mobility scenario, most queries are issued by the citizen systems. They target either the transport system, e.g. in order to compute route information, or other citizen systems, e.g. in order to find collocated routes or to determine whether two friends are in the same bus. Since some of the latter type of queries may be continuous queries, the remote processing system must interpret them – as continuous queries are not supported directly on Constrained Computer Systems (CCS).

In order to provide advanced behavior-driven services, the citizen systems and the bus systems are used to collect data collaboratively. As described previously, for citizen systems, this requires each citizen to opt in to the data collection and sharing by configuring the appropriate privacy settings. For bus systems, such a configuration is not necessary since the collected data does not affect privacy. Once relevant data is collected at the bus system or the citizen system, it is reported to the transport system.

The transport system collects the data received from the bus systems and citizen systems. Furthermore, it stores and aggregates it for service optimization purposes. This should typically result in local inferences as the aggregations required for the mobility scenario do not require dynamic data that is not available locally.

In order to make the optimized services accessible to the citizens, the application running on citizen systems provides an intent-aware user interface. Using the behavior information gathered by the data acquisition framework, the intent-aware user interface can notify the citizen about important events and it can display relevant information at the right time. In cases where the required predictions for this are imprecise or not possible, the citizen may specify goals using a speech recognition engine that is part of the framework.

In order to fetch the information that is relevant for the citizen, the intent-aware user interface issues queries and performs local or distributed inferences using the query processor and application-specific code. For some distributed inferences, it is necessary to access the data gathered by citizen systems of other citizens that share this data.

In order to enable privacy-preserving sharing, the privacy preservation framework controls the access to the data stored on the citizen systems. The basis for this is a privacy policy that is initialized using the information from external services such as Facebook or Google. The privacy preservation framework retrieves the privacy-related information from these systems periodically in order to determine relationships between different citizens and to keep the initial policy up-to-date. However, it is noteworthy that citizens can manipulate this generated policy through the user interface in order to customize it to their needs.

6.2.2 Environmental Scenario

The environmental application scenario is related to the mobility scenario due to the sources of information that are used for data collection. Specifically, the architectural instantiation described in the following relies on the data being captured by the bus system and a mobile application. Consequently,

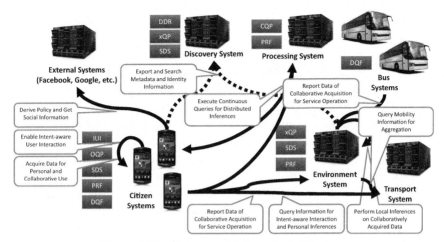

Figure 6.15 Environmental Scenario Architecture.

from an architectural perspective, the environmental scenario can be thought of as an extended version of the transport scenario. This is also clearly visible when comparing the instantiated architecture depicted in Figure 6.15 with the associated instantiation of the mobility scenario depicted in Figure 6.14. Nonetheless, we briefly describe both the deployment and the resulting interaction. For the sake of brevity, we refrain from revisiting the interactions with the transport system and focus on the environment system instead.

6.2.2.1 System Deployment

As depicted in Figure 6.15, the environmental scenario contains a number of computer systems that run various parts of the GAMBAS middleware as well as application-specific code that realizes the application functions. As indicated previously, a number of these systems are slight variations of the systems in the mobility scenario:

- **Citizen Systems:** Citizens are using their systems to gather information and to access services. For this, they rely on the same set of components as in the mobility scenario. However, the application-specific code has to be extended to accommodate the different usage.
- **Transport System:** Since some of the environmental use cases require transport-related information, the transport system of the mobility scenario is used to offer it. Specifically, the transport system is used to determine bus locations, which are required to provide the necessary context for environmental information.

- **Bus Systems:** Similar to the mobility scenario, the environmental scenario also makes use of embedded systems deployed in public buses in order to collect environmental data. The environmental data, however, will not be reported to the transport system but it will be reported to a new system – called the environmental system.
- **Discovery System:** Since the environmental scenario also requires distributed data processing, it is necessary to rely on the discovery system that manages the metadata and identity information.
- **Processing System:** Just like in the mobility scenario, a dedicated processing system is used to enable the citizen systems to run continuous queries against each other's devices. The processing system is equipped with a privacy framework and a continuous query processor.
- **External Systems:** The environmental scenario makes use of external systems to initialize the privacy policy. These systems are maintained by third parties, so no additional GAMBAS software can be installed on them and the necessary adapters are provided by the GAMBAS middleware app.

In addition to these systems which were also used for the mobility scenario, the architecture of the environmental scenario also introduces a new system:

- **Environment System:** Conceptually, the environment system is related to the transport system introduced in the mobility scenario as it manages and aggregates the environmental data reported by the bus and citizen systems. The main difference between the transport system and the environment system is the lack of a legacy data wrapper since the environment system does not have to tap into existing data sources. Other than that it, performs conceptually similar tasks such as data storage, aggregation and the computation of inferences.

6.2.2.2 System Interaction

In order to realize the different applications for the environmental scenario, the systems and their associated components have to interact with each other in a similar fashion as in the mobility scenario. The export of metadata and identity information is handled by the data storage components, the privacy framework and the data discovery registry. The searching is done transparently by the query processors and the privacy framework, which also create views on the semantic data storage of the citizen system, if necessary. The control for this is enabled by the policy generated by the privacy framework, which can be manually adjusted through the user interface.

Queries are issued by the citizen systems and the environmental system. They target either the transport system, e.g. in order to compute route information or other citizen systems. If a citizen system requires the execution of a continuous query, the remote processing system is used.

In order to provide advanced behavior-driven services, the citizen systems and the bus systems are used for collecting data collaboratively. The environment system collects the data received from the bus systems and citizen systems. Furthermore, it stores and aggregates the data, for example, to offer a pollution map, which can be used in conjunction with the transport system to compute alternative routes. This should typically result in local inferences at the environment system since the route information is mostly static and can be retrieved once for each computation.

In order to make the environmental services accessible to the citizens, the application running on citizen system provides an intent-aware user interface. In order to fetch the information that is relevant for the citizen, the intent-aware user interface issues queries and performs local or distributed inferences using the query processor and the application-specific code. This may entail distributed inferences, which are enabled by combining the continuous query processor on the processing system with the privacy frameworks on the citizen systems.

6.3 Application Components

As indicated by the application architecture, the implementation of the application scenarios entails a number of different components that are required to deliver the application functions. In the background, there are a number of application services that store and offer the data captured through sensing applications used by citizens or running in buses. In addition, there are background services that wrap legacy data coming from third-party data sources. Thus, in order to create a complete picture of the applications developed as part of GAMBAS, we first describe these application services. Thereafter, we outline the applications that we developed to capture the required data. On the basis of this description, we then describe the end-user applications for citizens as well as a set of innovative applications that feed the captured data back to the transit network operator.

6.3.1 Application Services

To power the mobility and environmental applications, we have developed a number of application-specific services using the GAMBAS middleware.

These services integrate with different data sources including data coming from EMT Madrid (incident feed, time tables, routes, etc.), open data provided by OpenStreetMap (addresses, geometry, etc.) and application-specific data (e.g. crowd-levels measured by the embedded applications running in vehicles). Although these services are conceptually backend services that are not directly visible to end users, the application services encompass frontends targeted at application developers and service administrators. In the following, we briefly walk through the different services and, where applicable, show a few screenshots of their frontends.

6.3.1.1 Tile Service
The tile service integrates with OpenStreetMap geometry data in order to generate images that are used to draw the map-based visualization. It supports multiple output formats and color schemes. The api has been designed to work with the Leaflet.js Javascript library, which is used consistently throughout the GAMBAS mobile applications. The screenshot shown in Figure 6.16(a) depicts a number of output options.

6.3.1.2 Incident Service
The incident service integrates with the EMT Madrid incident feed in order to provide incident information to the navigation application described later on.

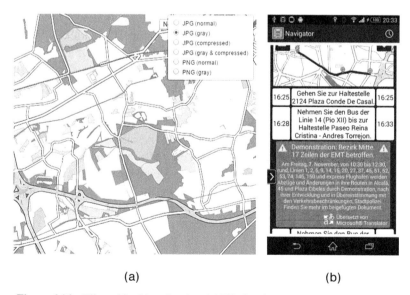

(a)　　　　　　　　　　(b)

Figure 6.16　Tile and Incident Service. (a) Tile Service and (b) Incident Service.

        (a)   (b)

Figure 6.17 Crowd and Routing Service. (a) Crowd Service and (b) Routing Service.

It is tightly integrated with the routing service in order to enable the output of route incidents for trips computed by the user. Since the EMT incident feed is only available in the Spanish language, the incident service has been integrated with Microsoft Translator, which provides machine translations into other languages supported by the mobile prototype applications. Figure 6.16(b) shows the resulting machine-translated output that is integrated into the routing result on the mobile app.

6.3.1.3 Crowd Service
The crowd service captures the crowd-level information generated by several buses in the city of Madrid. The captured data is then used by the routing service in order to provide crowd-level information as part of the routing result. To do this, the service aggregates the reports and assigns them to 15 minute timeslots, which are then used to drive the predictions. Figure 6.17(a) shows a sample crowd-level for one of these 15 minute timeslots.

6.3.1.4 Routing Service
The routing service (c.f. Figure 6.17(b)) integrates with the EMT GTFS data in order to compute crowd-aware routes, which are then used to power the navigation functions in the mobile application for citizens. In addition to transit routes using buses, it can also compute walking routes. If available, incident and crowd-level data will be returned directly as part of the routing result in order to minimize the amount of data that must be transferred between the mobile application and the service.

6.3.1.5 Network Service
The network service provides the mobile application for citizens with network-related information such as location and names of stops, routes of

(a)   (b)

Figure 6.18   Network and Timetable Service. (a) Network Service and (b) Timetable Service.

lines, etc. The resulting information is then used to visualize the route in the application. Using a tool, it is possible to extract the information from the network service and to ship it with the application. This is done in order to minimize the latency for displaying search results. Figure 6.18(a) shows the application which uses the built-in address database for auto-completion during place search.

6.3.1.6 Timetable Service
The timetable service provides the mobile navigation application with bus schedule information that is extracted from the GTFS information provided by EMT Madrid. Since the associated amount of GTFS data is too large to be processed directly on the device, this service takes care of extracting the relevant subsets based on a stop name and a calendar date. The mobile application then visualizes the output in a tabular form as depicted in Figure 6.18(b).

6.3.1.7 Geo Service
The geo service integrates with OpenStreetMap in order to resolve addresses into GPS coordinates. The service is used by the mobile applications to enable the user to search for addresses and to resolve GPS coordinates into addresses. The number of results returned by the service to the application is configurable in order to enable the optimization of applications for different criteria (i.e. bandwidth vs. flexibility). Figure 6.19(a) shows the output of the service on a map when searching for a particular address.

6.3.1.8 Log Service
The log service captures usage information generated by the mobile applications and enables the offline analysis of the user behavior for evaluation

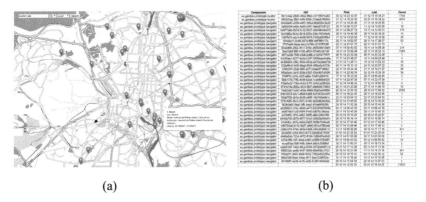

(a)   (b)

Figure 6.19   Geo and Log Service. (a) Geo Service and (b) Log Service.

purposes. It is based on a simple event abstraction that captures the reporting component, a generated user identifier, the time and the type of event as well as associated application-specific event data. A logging framework that has been integrated into the mobile applications is used to capture and synchronize the data produced by an application with the service. The captured events can be downloaded for later analysis. To do this, the service supports different queries based on event types, application components, dates, etc. In addition, the service can generate a report summary to track its internal status (e.g. which devices are uploading data, when devices have uploaded data and how much data has been captured already).

6.3.1.9 Noise Service

This service captures, aggregates and visualizes the noise information captured by a mobile sensing application. In addition, it can display the noise level for the captured locations. In order to avoid overloading clients, the data is aggregated inside the service before it is delivered to other applications. Figure 6.20 shows some sample noise data. The circles indicate locations where noise measurements are available. The circle color indicates the average noise-level at the location.

6.3.1.10 Environmental Service

The environmental service captures the environmental information gathered through measurements taken by sensors located in various buses that are driving through the city of Madrid. Thereby, the service associates the measurement with the real-time location of the bus. The service is equipped with

Figure 6.20 Noise Service.

Figure 6.21 Environment Service.

a simple user interface so that the individual measurments can be displayed. Thereby, it is possible to filter the measurements based on the sensor type as shown in Figure 6.21.

6.3.2 Sensing Applications

To provide data for the end-user applications, we have developed a number of sensing applications that target environmental information (noise, CO_2-level, pollen-levels, etc.) and transit information (i.e. crowd-levels of buses). As indicated in Section 6.2, this data is captured partly by mobile applications running on the devices of end-users and partly by embedded applications that are integrated into the buses that are operating in the city of Madrid. In the following, we briefly describe these sensing applications.

6.3.2.1 Noise-level Mobile App

To measure the noise profile at different points in the city, we extended the GAMBAS Locator application described in Section 6.1.4.3 with support for crowd-sensing. To do this, we integrated a data acquisition configuration that captures the sound profile using as the average frequency vector described in Chapter 3 and the sound pressure level. A user that wants to participate can activate the periodic background capturing of the sound profile through a settings screen. When activated, the sound profile is stored locally whenever the user's location is computed. As a result, the user can then visualize the the daily noise exposure as shown in Figure 6.22.

In addition to locally storing the information, the user can opt-in to crowd sensing. If the user enables crowd sensing, the sound profile and noise level will be uploaded to the noise service together with the user's current location and measurement time. Given a larger number of participants, the measurements can be used to create a picture of the noise profile of a city.

6.3.2.2 Pollution-level System

To capture pollution information in the city, we equipped a small number of buses with an environmental sensor as shown in Figure 6.23. Using an

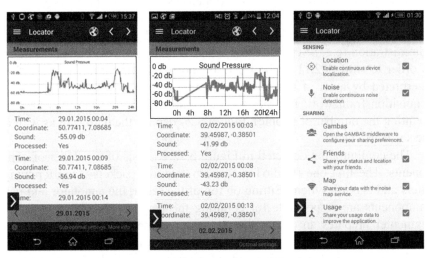

Figure 6.22   Noise-level Crowd-Sensing.

Figure 6.23 Embedded Sensors.

application embedded into the existing ICT infrastructure of the bus, we were using this deployment to continuously capture sensor readings while the bus was operating. The captured readings were then transmitted to the environmental service where the sensor data was stored together with the real-time GPS position of the bus.

6.3.2.3 Crowd-level System

One of the innovative functions of the mobile navigation app described in the following is to provide users with real-time and predicted crowd-level information about the vehicles on different routes. To capture this crowd-level information, we developed an embedded system and integrated into several buses [HIW+14]. The system consists of a TP-Link 3020, which is equipped with a Linux-based operating system (OpenWRT) running the JamVM virtual machine. Several operating system services have been specifically configured to enable a simple installation (e.g. DHCP, NTP) and to support remote administration (e.g. SSH, AutoSSH). The system uses pcaplib and tcpdump in order to sniff 802.11 probe requests and beacon frames. These are then interpreted by a set of components running on top of the GAMBAS data acquisition framework in order to determine the crowd-level of a bus (by counting the number of people). Figure 6.24 depicts the hardware as well as the software stack.

The configuration depicted in Figure 6.25 consists of a number of components. The first one (RadioTap Sensor) captures packets using tcpdump, filters and classifies them. Sitting on top of the senor, the annotator and gate components are responsible for counting the persons. Finally, the reading segmenter prepares an output file to be transmitted to a server at regular time intervals. These uploads are then performed asynchronously using the reading uploader. The reading uploader interacts with the crowd service, described previously, that stores the crowd-levels and makes them available

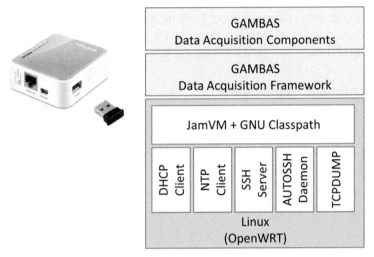

Figure 6.24 Embedded Crowd-Level Detection Application.

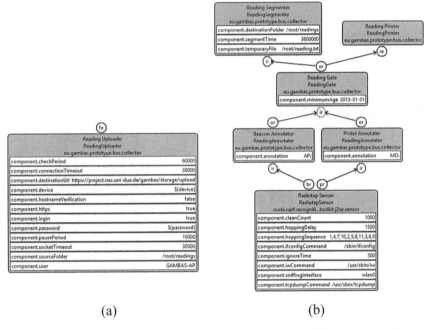

(a)   (b)

Figure 6.25 Crowd-Level Detection Configuration.

to the mobile application. In order to mitigate potential privacy issues, however, the collected data is anonymized by removing any personal identifiable information (i.e. device MAC addresses) before it is uploaded.

6.3.3 End-user Applications

The end-user applications provide regular citizens with the ability to access the information captured through the sensing applications and managed by the GAMBAS application services. In the following, we briefly outline the two end-user applications that have been developed for the mobility and the environmental scenario.

6.3.3.1 Navigation App

For the mobility scenario, we developed a mobile application for Android. Since the mobility-related application services are integrating data from the public bus network of the city of Madrid, we called this application **Madrid Navigator.**

The Madrid Navigator is a maps and navigation application that is conceptually similar to other modern navigation applications for mobile phones. As depicted in Figure 6.26, it provides users with a map of their environment and allows them to search for places and bus stops. Using the voice control

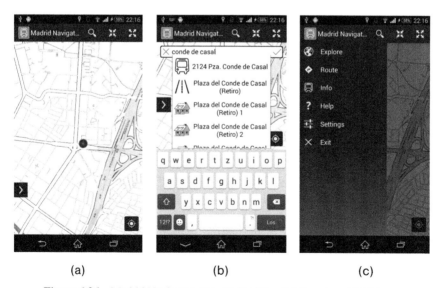

| (a) | (b) | (c) |

Figure 6.26   Madrid Navigator App. (a) Position, (b) Search and (c) Menu.

components described in Chapter 3, users can not only search for places via text input but also through speech input. Using different icons, the Madrid Navigator categorizes search results into cities, streets, buildings and bus stops. Depending on the category, the Madrid Navigator can show additional information such as bus routes going through a particular stop or timetables.

In addition to retrieving additional information, the search results can also be used to compute routes. To do this, a user can simply pick any place on the map and tap on a route button. Alternatively, the user can enter a source and a destination address or GPS coordinate into the routing screen shown in Figure 6.28. On this screen, a user can also adjust different parameters such as the desired arrival or departure time and specify the desired modality (e.g. on foot or by bus). When the user is satisfied and starts the computation, the specified parameters are transmitted to a GAMBAS service that computes one or more route alternatives. Once the routes have been computed, they are visualized in a list of route summaries. The user can inspect this list and get additional information by tapping on one of the summaries.

As shown in Figure 6.27, the detailed view not only shows information about the sequence of actions on the route, but it also depicts crowd-level information on the specific bus that is proposed. This allows users to compare consecutive trips on the same route with respect to the expected crowdedness

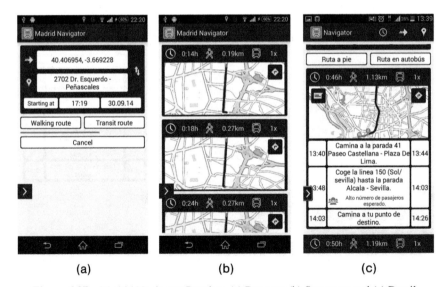

Figure 6.27   Madrid Navigator Routing. (a) Request, (b) Summary and (c) Detail.

of the bus. To determine this information, we use the embedded sensing application described previously, which we deployed in several buses running through the city. Using this embedded application, we collect real-time information about the number of passengers on board of the buses. The resulting information is then processed in order to compute predictions for other buses, which are not equipped with the crowd-level sensing application. The resulting predictions are then fed back into the routing service such that they can be used (a) to guide routing decisions and (b) to inform the users.

Once a user has decided to follow a particular route proposal, the user can start a navigation session for the route. During this session, the GAMBAS middleware can automatically share the user's intended destination with the transit network operator. As explained later on, this allows the operator to detect routes that are going to be in high demand in the near future. Thereby, the user's identity is hidden from the operator. During navigation session, the user is supported through step-by-step instructions as shown in Figure 6.28.

The step-by-step instructions implement the concept of micro-navigation described in [FKR+14]. The idea behind micro-navigation is to optimally support the user's information needs during the usage of public transportation. For this, the app must provide the right pieces of information at the time when they are needed. To do this, the application usage text messages

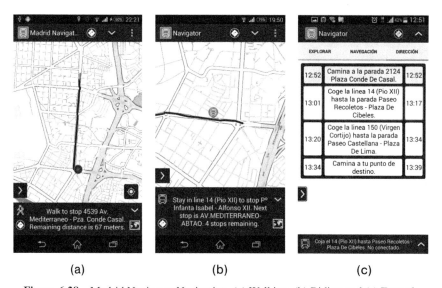

Figure 6.28 Madrid Navigator Navigation. (a) Walking, (b) Riding and (c) Textual.

that are shown at the bottom of the screen at all times. In addition, the application provides (optional) voice output using text-to-speech. To generate instructions, the application uses the GAMBAS data acquisition framework to tap into the sensors and information provided by the bus. To do this, the application automatically connects to the Wi-Fi network available in every bus operated by EMT Madrid and connects to the internal information system to determine the location and route of the bus. This information is then used to generate messages that correspond exactly to the user's context. For example, the app will notify the user to get off the bus shortly before it arrives at the correct stop. Similarly, if the user has taken the wrong bus, the app will immediately inform the user and propose a corrective action (e.g. to re-plan the route or to exit the bus at the next stop).

As shown in Figure 6.29, the app also enables users to directly access the bus information whenever they are traveling. This allows them to get real-time information about expected arrival times, even if they are not using micro-navigation. In addition, the application also integrates with the incident feed provided by the bus operator. This incident feed describes changes to schedules, e.g. due to demonstrations in the city center or traffic accidents. Thereby, the incidents are directly integrated into the routing results as well

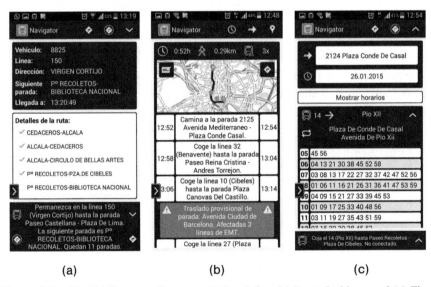

Figure 6.29 Madrid Navigator Features. (a) Bus Infos, (b) Route Incidents and (c) Time Table.

as the timetable information that can be fetched for different stops. In addition to incidents, the timetable information also includes real-time information for buses that are departing within the next 20 minutes. To do this, the application integrates with a real-time service provided by EMT Madrid through the GAMBAS middleware.

6.3.3.2 Environmental Map

For the environmental scenario, we have developed a web-based application that enables end-users to inspect the state of the environment. This state is captured through measurements of pollutants that are acquired via the sensor deployment in buses and the noise-level measurements of the mobile noise-sensing application. For this, the environmental map application integrates with the GAMBAS noise service and the environmental service, described previously. After retrieving the data from them, the application applies the following data aggregation approach to create a visually appealing data representation:

1. Values corresponding to measurements at certain locations are clustered.
2. Based on the clusters, we identify the Voronoi partitions to define the area of the cluster.
3. Using Delaunay triangulation, we find adjacent areas to interpolate missing data.
4. Finally, we perform hexagonal binning in order to represent the result.

The resulting hexagonal visualization is then added to an overlay of tiles computed with the tile service, which results in the final result shown in Figure 6.30. Thus, using the web-based application, a user can simply move

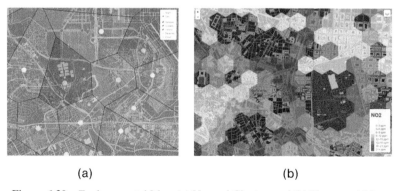

(a)   (b)

Figure 6.30   Environmental Map. (a) Voronoi Clusters and (b) Hexagonal Map.

the map to a specific location in the city and then view the different sensor readings in a manner that is easy to understand.

6.3.4 Operator Applications

In addition to the application services, sensing applications and mobile applications, we have also developed a number of applications that are not targeting the citizens. Instead, they are targeted towards the transit network operator, which, in our specific case, is EMT Madrid. The operator applications are aggregating the information collected through the crowd-level sensors and the mobile applications in order to help the operators to understand the current transit network usage. This understanding can then be used to optimize the network, possibly in real time, e.g. by dispatching additional buses or issuing route warnings, etc. In the following, we briefly outline these services.

6.3.4.1 Congestion Notifications

At the EMT Madrid headquarter, there are operators that control all the operations related to the bus network management. Crowd-level detection provides an estimation of the bus occupancy. A bus is considered as "congested" when a threshold of 85% of its capacity is exceeded. When the embedded application on the bus detects that a bus is getting congested, it generates a notification to signal this to the operator. If 2/3 of the vehicles within a route are congested, then the operator receives another notification that signals the congestion in the route. These alarms and notifications have been incorporated to the management system in a way that they can be visualized in the same graphical user interface that EMT is currently using. Figure 6.31 shows how the operator that is managing a route is notified when the threshold level is exceeded, by displaying an "Ocupacion LLENO" message (full occupation) and in red, the message "Ruta atocha-misericordia congestionada" (Atocha-misericordia route congested).

6.3.4.2 Demand Notifications

As described previously, the most demanded routes by the Madrid Navigator users are detected based on the usage of the navigation functionality. The currently used destinations during navigation are stored in the demand service. Once a certain number of destinations located in a certain area, for a given period of time, are reached, then that area can be categorized as a high-demanded destination zone. As a result, it will be shown to the bus

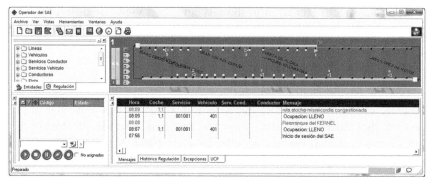

Figure 6.31 Congestion Notifications.

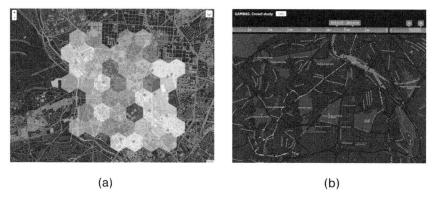

(a)   (b)

Figure 6.32 Demand Notifcations and Occupancy Analysis. (a) Demand Notification and (b) Occupancy Analysis.

network operators who can use this information to detect a massive event such as a concert or a demonstration. Based on this, the operator can decide whether to reinforce the related bus lines covering that area or not. The information is offered to the bus operator in a map by using the hexagonal binning representation. The different hexagonal areas allow the operator to visualize the most demanded destinations in a quick and simple manner, as shown in Figure 6.32(a).

6.3.4.3 Occupancy Analysis

Crowd-level measurements are received in real time and stored in a data storage for offline analysis. For this storage service, we developed an operator tool to display the real-time and historical bus occupancy. Using this tool, the bus network operator is able to visualize occupancy information in a geo-located manner for a selected bus line. The viewer is implemented as a web application to visualize the buses location integrated with a map. The colors in the different routes are showing the crowd-level data at a specific time: low-crowded (green), medium-occupied (orange) and congested (red), in the same way as this information is shown in the mobile app.

6.4 Application Evaluation

During the course of the development of the GAMBAS middleware, we deployed all application services and sensing components. In addition, we performed a large-scale deployment of the Madrid Navigator navigation application. For the operator applications and the pollution map, we performed only internal testing with a closed user group. During the internal testing of the environmental applications, we found that the pollutant sensing system in the bus was not able to collect meaningful data. After an analysis and several rounds of discussions with the hardware manufacturer of the pollution sensor, we stopped the further roll-out of the system due to the unreliability of the sensor readings. As a consequence, the evaluation results described in the following are centered around the mobility scenario and the navigation app in particular.

To evaluate the Madrid Navigator navigation app, we distributed it through the Android market in order to make it available to interested users and application developers. During the evaluation period, the application was downloaded more than 1000 times and used by both an internal group of testers and actual citizens that were not related to GAMBAS. From this deployment, we collected a significant amount of feedback both implicit (through the app usage) and explicit (through in-app questions and a feedback form). In the following, we briefly describe the application functionality and the results gathered during the deployment.

In order to detect issues and to improve the app during the deployment, we instrumented it with logging code. If a user gave his explicit consent as shown in Figure 6.33, we uploaded and analyzed the logs using the logging service described previously. In addition to implicit feedback, we also offered two ways to provide explicit feedback. First, we integrated a feedback form into the application and second, we used pop-up dialogs to ask user's about their current experience. For this, we implemented a regular 5-star rating dialog shown in Figure 6.33. Using the in-app questions, the users collectively generated 350 responses to different questions. Each of these questions could

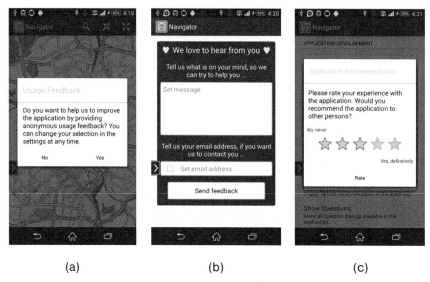

(a)   (b)   (c)

Figure 6.33   Madrid Navigator Feedback. (a) Implicit, (b) Form and (c) Question.

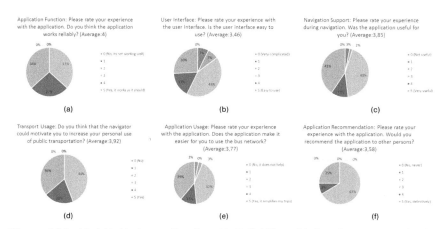

(a)   (b)   (c)

(d)   (e)   (f)

Figure 6.34   Madrid Navigator Results. (a) Reliability, (b) Interface, (c) Navigation, (d) Motivation, (e) Usage and (f) Recommendation.

have been rated between 0 and 5 stars. The responses to each question are shown in Figure 6.34. In the following, we briefly discuss the results.

To determine whether the application worked as expected on the broad number of devices of the users, we asked the users to provide a rating with respect to reliability. As depicted above, 36% of the users gave a 5-star rating (works as it should), 27% of the users gave a 4-star rating and 37% of the users gave a 3-star rating resulting in an average rating of 4 (out of 5). Consequently, we think that the mobile application was working well in many cases as none of the users gave a rating that was worse than 3 stars.

The second question that we posed to the users was to rate the overall usability of the user interface between easy-to-use (5 stars) and very complicated (0 stars). With 43%, the majority of users thought that the interface is neither easy nor complicated to use. Another 43% assigned a 4 or 5 star rating marking the interface clearly as easy-to-use. However, on the negative side, 14% of the users thought that the interface was rather complicated. We speculated that this could be due to issues on devices that have a small screen, which could result in usability issues with the map-based visualizations (e.g. small icons, etc.). However, we were not able to prove this assumption.

In addition to crowd-level and incident-aware routing, one of the core features of the GAMBAS Madrid Navigator is the application of context-awareness to enable intent-aware navigation instructions. Thus, in order to evaluate the usefulness of this feature, we asked the users whether they consider the navigation to be useful. Here, the overwhelming majority of users (95%) is rating the application with a 3 star or higher rating. 41% are rating the application even with the maximum rating resulting in an average of 3.85 stars. This clearly shows that a) the navigation was working reliable and b) the idea of micro-navigation was clearly considered to be useful.

In order to determine the impact of the Madrid Navigator on the user's transport behavior, we asked whether they think that the application could motivate them to use more public transportation. Here, the answer is again rather positive since 36% of the users completely agree that the application could motivate them and another 20% rather agrees, which results in an average rating of 3.92 stars. Consequently, we argue that navigation applications like the Madrid Navigator that employ context- and intent-awareness can be a benefit for transport network operators.

To determine whether the application actually helps users during their trips, we asked whether the application makes it easier for them to use the bus network. Again, the overall results were rather positive since 39% of the users stated that it would simplify their trips at least somewhat and only 6% answered that the application would not help. Thus, the overwhelming majority of users providing detailed and precise navigation instructions by means of context recognition at the right point in time can simplify their bus trips.

Finally, to gather the users overall impression on the Madrid Navigator, we asked them whether they would recommend the application to other users. Just like with previous questions on the reliability, usability and helpfulness of the application, the explicit user feedback reveals a rather positive result. With an average of 3.58 stars, the users are either undecided or would recommend the application.

In summary, these results are a clear indication for the maturity and usefulness of the navigation application. Given the fact that the implementation of the Madrid Navigator and all of its background services was leveraging the GAMBAS middleware, this also demonstrates the applicability of the abstractions provided by it. Thereby, it is important to stress that, in contrast to many other research projects, the tests were performed under realistic conditions with a large number of users that were not affiliated with the GAMBAS project.

7

Conclusion

The GAMBAS middleware encompasses several subsystems covering an adaptive data acquisition, interoperable data modeling and distributed processing, automated privacy preservation as well as associated user interfaces. As described in Chapter 3, Chapter 4 and Chapter 5, these subsystems can be further split into concepts, frameworks, mechanisms and protocols. In the following, we briefly revist their functions and highlight their technical innovations.

Given that the GAMBAS middleware aims at supporting behavior-driven services, the data acquisition framework is clearly one of the fundamental building blocks of the GAMBAS middleware. Conceptually, the framework is responsible for context recognition on personal mobile devices including smart phones, PDAs and laptops. The framework supports various platforms including Android, Windows and Linux. It follows a multi-stage approach, which enables the development of context recognition applications from generic components that can be executed in an energy-efficient manner. To do this, the data acquisiton framework leverages a component abstraction to foster genericity and a state machine abstraction to enable the energy-efficient execution of complex context recogniton logic. Both these subsystems have been extensively used in development of various applications and technical demonstrations of GAMBAS. Some of the examples include the recognition of the user's context during a multimodal trip in order to support micro-navigation or the detection of noise and crowd levels to improve the user experience while traveling.

In order to make the acquired data usable by different applications, the GAMBAS middleware introduces interoperable data representations that follow the linked open data principles and are based on semantic web technologies. The GAMBAS ontology not only enables different applications to leverage the same data, but it is also used internally by the middleware, for example, to model users and privacy policies. It is also noteworthy to mention that the ontology, that is accompanying the GAMBAS middleware, is not trying to reinvent the necessary concepts. Instead, it integrates a large number of ontologies that are already actively used. This increases the compatibility and simplifies the application development. On top of the interoperable data representations, the GAMBAS middleware introduces a dynamic data processing system that features a semantic-based auto-discovery powered by an associated linked open-data infrastructure. This infrastructure leverages a dynamic data registry to make data available across arbitrary applications, and it features data storages that can be queried locally and remotely. Using the efficient implementation of semantic data storages, it is possible to use standard query languages for semantic data even on resource-poor mobile devices while maintaining a query performance that is suitable for complex applications such as the mobile navigation application described in Chapter 6. Using specifically designed language extensions such as CQELS, the GAMBAS middleware not only supports queries on static data, but instead, it also allows evaluation of continuous queries over dynamic data streams. However, since this requires a higher amount of processing power, this support is not integrated directly into mobile devices. Instead, the GAMBAS middleware uses a distributed query processing architecture that offloads the effort to more powerful systems.

As a result of the automated acquisition of context information and the distributed processing of context information enabled by the GAMBAS middleware, security and privacy are becoming key issues that must be considered. For this reason, the GAMBAS middleware encompasses mechanisms and protocols to automate the preservation of the user's privacy as far as possible. In this context, it is worth pointing out that the GAMBAS privacy preservation framework goes well beyond encrypted communication by managing the access to the user's data on the basis of the user's privacy policy. To do so, it integrates with all other system components including the data acquisition framwork, dynamic data registry and the semantic data storages running on devices of the user's and services deployed on the Internet. To implement access control on top of authenticated communication, the privacy framework allows users to automatically bootstrap the required encryption keys through popular online services such as Facebook. This not only minimizes the friction of secure data sharing, but it also enables secure peer-based (i.e. server-less) sharing of data between user devices without any manual configuration. Similarly, to minimize the user effort for setting up privacy policies, their generation can be (partially) automated through these services as well. Towards this end, the privacy framework encompasses a policy generator that interprets the sharing behavior of a user to derive a suitable policy generation.

Together, these concepts, frameworks, protocols and mechanisms provide a generic structure that simplifies the development of behavior-driven services. This has been successfully demonstrated by the large number of applications and services that have been built using the GAMBAS middleware during its development. The applications implement several innovative features that are based on the user's behavior. This includes the automated capturing of user-specific information (e.g. intended trip destinations, noise exposure, etc.) as well as the privacy-preserving sharing of derived information (e.g. crowd levels, high-demanded routes, noise pollution in the city, etc.). The resulting data can be made available to service operators such as the bus network operators from EMT Madrid, which allows them to optimize their services, e.g. by dispatching more buses when high-demanded routes or destinations are detected.

Over the course of the project, the GAMBAS middleware has been made publicly available to third-party developers. The full source code of GAMBAS is available via a public Maven repository that can be reached through the project website. The source distribution includes tutorials and example applications to showcase and demonstrate the use of the middleware to simplify the development of applications leveraging behavior-driven services. In addition, the distribution also includes binaries in the form of a software development kit that is packaged to support application development on a broad range of different platforms.

From an academic perspective, the development of the GAMBAS middleware has resulted in a significant amount of research contributions beyond the state of the art. During the 3-year-long development of the GAMBAS middleaware and its applications, the members of the GAMBAS consortium published specific concepts, algorithms and evaluations in more than 25 papers and articles in academic conferences and journals with high visibility. Furthermore, the consoritum organized 5 different stakeholder workshops that shaped the design of the middleware significantly. Finally, the consortium demonstrated the GAMBAS technology and its applications at 3 different industrial events in order to disseminate the research beyond the academic sector.

In addition to publications, the availability of the GAMBAS middleware has resulted in a considerable pickup of the underlying implementations and concepts through other research projects. For example, the SIMON Project[1] has reused most of the mobility-related services to implement a mobile application that provides mobility support for disabled and elderly persons in 4 major European cities. The SmartKYE Project[2] has reused the highly configurable component-based approach to data acquisition and processing provided by the data acquisition framework of the GAMBAS middleware in their energy-management infrastructure. Finally, the BESOS Project[3] has reused concepts from the privacy framework and the SmartAction Project[4] has reused GAMBAS for a joint IOT-middleware demonstration.

Given the current computing landscape with mostly centralized IoT infrastructures, we hope that this book will further strengthen the pickup of the approaches, concepts and technology developed and validated by the GAMBAS middleware and its applications.

---

[1]SIMON Project Homepage: http://simon-project.eu/
[2]SmartKye Project Homepage: http://smartkye.eu/
[3]BESOS Project Homepage: http://besos-project.eu/
[4]SmartAction Project Homepage: http://www.smart-action.eu/

Bibliography

[3PC12] 3PC. 3PC Project, Project Homepage. http://www.3pc.info, 2012. Accessed: April 2012.

[ACDCdVS08] C. A. Ardagna, M. Cremonini, S. De Capitani di Vimercati, and P. Samarati. A privacy-aware access control system. *J. Comput. Secur.*, 16(4):369–397, December 2008.

[AFRS11] Darko Anicic, Paul Fodor, Sebastian Rudolph, and Nenad Stojanovic. Ep-sparql: A unified language for event processing and stream reasoning. In *Proceedings of the 20th International Conference on World Wide Web*, WWW '11, pages 635–644, New York, NY, USA, 2011. ACM.

[AHIM13] W. Apolinarski, M. Handte, M. U. Iqbal, and P. J. Marrn. Piggy-backed key exchange using online services (pike). In *2013 IEEE International Conference on Pervasive Computing and Communications Workshops (PERCOM Workshops)*, pages 309–311, March 2013.

[AHM12] W. Apolinarski, M. Handte, and P. J. Marrn. An approach for secure role assignment. In *2012 Eighth International Conference on Intelligent Environments*, pages 34–41, June 2012.

[ALL12] ALLOW. ALLOW FET Proactive, Project Homepage. http://www.allow-project.eu, 2012. Accessed: April 2012.

[Apa13] Apache Foundation. Apache jena homepage. http://jena.apache.org/, 2013. Accessed: May 2013.

[BAL08] D. Bannach, O. Amft, and P. Lukowicz. Rapid prototyping of activity recognition applications. *IEEE Pervasive Computing*, 7(2):22–31, April 2008.

[Bar05] Jakob E. Bardram. The java context awareness framework (jcaf) – a service infrastructure and programming framework for context-aware applications. In Hans W. Gellersen, Roy Want, and Albrecht Schmidt, editors, *Pervasive Computing*, pages 98–115, Berlin, Heidelberg, 2005. Springer Berlin Heidelberg.

[BBCG10] Davide Francesco Barbieri, Daniele Braga, Stefano Ceri, and Michael Grossniklaus. An execution environment for c-sparql queries. In *Proceedings of the 13th International Conference on Extending Database Technology*, EDBT '10, pages 441–452, New York, NY, USA, 2010. ACM.

[BC09] Xuan Bao and Romit Roy Choudhury. Vupoints: Collaborative sensing and video recording through mobile phones. In *Proceedings of the 1st ACM Workshop on Networking, Systems, and Applications for Mobile Handhelds*, MobiHeld '09, pages 7–12, New York, NY, USA, 2009. ACM.

[BEH⁺06] J. Burke, D. Estrin, M. Hansen, A. Parker, N. Ramanathan, S. Reddy, and M. B. Srivastava. Participatory sensing. In *Workshop on World-Sensor-Web (WSW06): Mobile Device Centric Sensor Networks and Applications*, pages 117–134, 2006.

[BFL⁺07] Eric Bouillet, Mark Feblowitz, Zhen Liu, Anand Ranganathan, Anton Riabov, and Fan Ye. A semantics-based middleware for utilizing heterogeneous sensor networks. In James Aspnes, Christian Scheideler, Anish Arora, and Samuel Madden, editors, *Distributed Computing in Sensor Systems*, pages 174–188, Berlin, Heidelberg, 2007. Springer Berlin Heidelberg.

[BGJ08] Andre Bolles, Marco Grawunder, and Jonas Jacobi. Streaming sparql – extending sparql to process data streams. In Sean Bechhofer, Manfred Hauswirth, Jörg Hoffmann, and Manolis Koubarakis, editors, *The Semantic Web: Research and Applications*, pages 448–462, Berlin, Heidelberg, 2008. Springer Berlin Heidelberg.

[BHBL09] Christian Bizer, Tom Heath, and Tim Berners-Lee. Linked data – the story so far. *International Journal on Semantic Web Information Systems*, 5:1–22, 2009.

[BM10] Ana M. Bernardos and José R. Madrazo, Evaand Casar. An embeddable fusion framework to manage context information in mobile devices.In Emilio Corchado, Manuel Graña Romay, and Alexandre Manhaes Savio, editors, *Hybrid Artificial Intelligence Systems*, pages 468–477, Berlin, Heidelberg, 2010. Springer Berlin Heidelberg.

[BS04] A. R. Beresford and F. Stajano. Mix zones: user privacy in location-aware services. In *IEEE Annual Conference on Pervasive Computing and Communications Workshops, 2004. Proceedings of the Second*, pages 127–131, March 2004.

[BSGR03] C. Becker, G. Schiele, H. Gubbels, and K. Rothermel. Base – a micro-broker-based middleware for pervasive computing. In *Proceedings of the First IEEE International Conference on Pervasive Computing and Communications, 2003. (PerCom 2003).*, pages 443–451, March 2003.

[CBSG12] Heng-Tze Cheng, Senaka Buthpitiya, Feng-Tso Sun, and Martin Griss. Omnisense: A collaborative sensing framework for user context recognition using mobile phones. In *Elevent international workshop on Mobile Computing Systems and Applications Hotmobile*, 2012.

[CCI89] CCITT. The directory-authentication framework. Recommendation X.509, 1989.

[CEL+06] Andrew T. Campbell, Shane B. Eisenman, Nicholas D. Lane, Emiliano Miluzzo, and Ronald A. Peterson. People-centric urban sensing. In *Proceedings of the 2nd Annual International Workshop on Wireless Internet*, WICON '06, New York, NY, USA, 2006. ACM.

[CK00] Guanling Chen and David Kotz. A survey of context-aware mobile computing research. Technical report, Dartmouth College, Hanover, NH, USA, 2000.

[CNR12] CNR. DOLCE: a descriptive ontology for linguistic and cognitive engineering. http://www.loa.istc.cnr.it/DOLCE.html, 2012. Accessed: May 2012.

[CYCS12] J. Carrapetta, N. Youdale, A. Chow, and V. Sivaraman. Haze Watch Project, Project Homepage. http://www.pollution.ee.unsw.edu.au, 2012. Accessed: April 2012.

[DER12] DERI. Privacy preference ontology. http://vocab.deri.ie/ppo, 2012. Accessed: May 2012.

[DHH07] Anusuriya Devaraju, Simon Hoh, and Michael Hartley. A context gathering framework for context-aware mobile solutions. In *Proceedings of the 4th International Conference on Mobile Technology, Applications, and Systems and the 1st International Symposium on Computer Human Interaction in Mobile Technology*, Mobility '07, pages 39–46, New York, NY, USA, 2007. ACM.

[EBBS07] D. Evans, A. R. Beresford, T. Burbridge, and A. Soppera. Context-derived pseudonyms for protection of privacy in transport middleware and applications. In *Pervasive Computing and Communications Workshops, 2007. PerCom Workshops '07. Fifth Annual IEEE International Conference on*, pages 395–400, March 2007.

[EML+07] S. B. Eisenman, E. Miluzzo, N. D. Lane, R. A. Peterson, G-S. Ahn, and A. T. Campbell. The bikenet mobile sensing system for cyclist experience mapping. In *Proceedings of the 5th International Conference on Embedded Networked Sensor Systems*, SenSys '07, pages 87–101, New York, NY, USA, 2007. ACM.

[EML⁺10] Shane B. Eisenman, Emiliano Miluzzo, Nicholas D. Lane, Ronald A. Peterson, Gahng-Seop Ahn, and Andrew T. Campbell. Bikenet: A mobile sensing system for cyclist experience mapping. *ACM Trans. Sen. Netw.*, 6(1):6:1–6:39, January 2010.

[FKR⁺14] Stefan Foell, Gerd Kortuem, Reza Rawassizadeh, Marcus Handte, Umer Iqbal, and Pedro Marrón. Micro-navigation for urban bus passengers: Using the internet of things to improve the public transport experience. In *Proceedings of the First International Conference on IoT in Urban Space*, URB-IOT '14, pages 1–6, ICST, Brussels, Belgium, Belgium, 2014. ICST (Institute for Computer Sciences, Social-Informatics and Telecommunications Engineering).

[FL10] Lujun Fang and Kristen LeFevre. Privacy wizards for social networking sites. In *Proceedings of the 19th International Conference on World Wide Web*, WWW '10, pages 351–360, New York, NY, USA, 2010. ACM.

[FOA12] FOAF. Foaf vocabulary specification. http://xmlns.com/foaf/spec/, 2012. Accessed: May 2012.

[FSH12] Julien Freudiger, Reza Shokri, and Jean-Pierre Hubaux. Evaluating the privacy risk of location-based services. In George Danezis, editor, *Financial Cryptography and Data Security*, pages 31–46, Berlin, Heidelberg, 2012. Springer Berlin Heidelberg.

[GG03] Marco Gruteser and Dirk Grunwald. Anonymous usage of location-based services through spatial and temporal cloaking. In *Proceedings of the 1st International Conference on Mobile Systems, Applications and Services*, MobiSys '03, pages 31–42, New York, NY, USA, 2003. ACM.

[GHJV95] Erich Gamma, Richard Helm, Ralph Johnson, and John Vlissides. *Design Patterns: Elements of Reusable Object-oriented Software*. Addison-Wesley Longman Publishing Co., Inc., Boston, MA, USA, 1995.

[GHV07] C. Gutierrez, C. A. Hurtado, and A. Vaisman. Introducing time into rdf. *IEEE Transactions on Knowledge and Data Engineering*, 19(2):207–218, Feb 2007.

[GJAS06] Raghu K. Ganti, Praveen Jayachandran, Tarek F. Abdelzaher, and John A. Stankovic. Satire: A software architecture for smart attire. In *Proceedings of the 4th International Conference on Mobile Systems,*

Applications and Services, MobiSys '06, pages 110–123, New York, NY, USA, 2006. ACM.

[Goo12] GoodRelations. Goodrelations language reference. http://purl. org/goodrelations/v1, 2012. Accessed: May 2012.

[GSSS02] David Garlan, Dan Siewiorek, Asim Smailagic, and Peter Steenkiste. Project aura: Toward distraction-free pervasive computing. *IEEE Pervasive Computing*, 1(2):22–31, April 2002.

[Han09] M. Handte. System-support for adaptive pervasive applications, PhD Thesis. Universität Stuttgart, 2009.

[HH10] Michael Haslgrübler and Clemens Holzmann. Darsens: A framework for distributed activity recognition from body-worn sensors. In *Proceedings of the Fifth International Conference on Body Area Networks*, BodyNets '10, pages 240–246, New York, NY, USA, 2010. ACM.

[HIW+14] Marcus Handte, Muhammad Umer Iqbal, Stephan Wagner, Wolfgang Apolinarski, Pedro José Marrón, Eva Maria Muñoz Navarro, Santiago Martinez, Sara Izquierdo Barthelemy, and Mario González Fernández. Crowd density estimation for public transport vehicles. In *EDBT/ICDT Workshops*, 2014.

[HKL+99] Fritz Hohl, Uwe Kubach, Alexander Leonhardi, Kurt Rothermel, and Markus Schwehm. Next century challenges: Nexus—an open global infrastructure for spatial-aware applications. In *Proceedings of the 5th Annual ACM/IEEE International Conference on Mobile Computing and Networking*, MobiCom '99, pages 249–255, New York, NY, USA, 1999. ACM.

[HM08] D. Henrici and P. Mller. Providing security and privacy in rfid systems using triggered hash chains. In *2008 Sixth Annual IEEE International Conference on Pervasive Computing and Communications (PerCom)*, pages 50–59, March 2008.

[HV09] Dijiang Huang and Mayank Verma. Aspe: Attribute-based secure policy enforcement in vehicular ad hoc networks. *Ad Hoc Netw.*, 7(8):1526–1535, November 2009.

[HWS+10] M. Handte, S. Wagner, G. Schiele, C. Becker, and P. J. Marron. The base plug-in architecture - composable communication support for pervasive systems. In *7th ACM International Conference on Pervasive Services*, July 2010.

[IET13] IETF. Ietf geopriv charter. http://datatracker.ietf.org/wg/ geopriv/charter/, 2013. Accessed: May 2013.

[IHW$^+$12] Muhammad Umer Iqbal, Marcus Handte, Stephan Wagner, Wolfgang Apolinarski, and Pedro José Marrón. Enabling energy-efficient context recognition with configuration folding. *2012 IEEE International Conference on Pervasive Computing and Communications*, pages 198–205, 2012.

[IP08] Vincenzo Iovino and Giuseppe Persiano. Hidden-vector encryption with groups of prime order. In Steven D. Galbraith and Kenneth G. Paterson, editors, *Pairing-Based Cryptography – Pairing 2008*, pages 75–88, Berlin, Heidelberg, 2008. Springer Berlin Heidelberg.

[JJFZ11] P. Jagtap, A. Joshi, T. Finin, and L. Zavala. Preserving privacy in context-aware systems. In *2011 IEEE Fifth International Conference on Semantic Computing*, pages 149–153, Sept 2011.

[JS10] S. Ji and D. Shin. An efficient garbage collection for flash memory-based virtual memory systems. *IEEE Transactions on Consumer Electronics*, 56(4):2355–2363, November 2010.

[Kal00] B. Kaliski. Pkcs 5: Password-based cryptography specification version 2.0. RFC 2898, September 2000.

[KLJ$^+$08] Seungwoo Kang, Jinwon Lee, Hyukjae Jang, Hyonik Lee, Youngki Lee, Souneil Park, Taiwoo Park, and Junehwa Song. Seemon: Scalable and energy-efficient context monitoring framework for sensor-rich mobile environments. In *Proceedings of the 6th International Conference on Mobile Systems, Applications, and Services*, MobiSys '08, pages 267–280, New York, NY, USA, 2008. ACM.

[KZX$^+$11] Matthew Keally, Gang Zhou, Guoliang Xing, Jianxin Wu, and Andrew Pyles. Pbn: Towards practical activity recognition using smartphone-based body sensor networks. In *Proceedings of the 9th ACM Conference on Embedded Networked Sensor Systems*, SenSys '11, pages 246–259, New York, NY, USA, 2011. ACM.

[LHLY09] Yinan Li, Bingsheng He, Qiong Luo, and Ke Yi. Tree indexing on flash disks. In *Proceedings of the 25th International Conference on Data Engineering, ICDE 2009, March 29 2009 - April 2 2009, Shanghai, China*, page 13031306, March 2009.

[Lin12] Linked Data. Linked Data, homepage. http://linkeddata.org, 2012. Accessed: May 2012.

[LLEC08] Nicholas D. Lane, Hong Lu, Shane B. Eisenman, and Andrew T. Campbell. Cooperative techniques supporting sensor-based people-centric inferencing. In Jadwiga Indulska, Donald J. Patterson, Tom Rodden, and Max Ott, editors, *Pervasive Computing*, pages 75–92, Berlin, Heidelberg, 2008. Springer Berlin Heidelberg.

[LNK⁺07] Sang-Won Lee, Gap-Joo Na, Jae-Myung Kim, Joo-Hyung Oh, and Sang-Woo Kim. Research issues in next generation dbms for mobile platforms. In *Proceedings of the 9th International Conference on Human Computer Interaction with Mobile Devices and Services*, MobileHCI '07, pages 457–461, New York, NY, USA, 2007. ACM.

[LPDTXPH11] Danh Le-Phuoc, Minh Dao-Tran, Josiane Xavier Parreira, and Manfred Hauswirth. A native and adaptive approach for unified processing of linked streams and linked data. In Lora Aroyo, Chris Welty, Harith Alani, Jamie Taylor, Abraham Bernstein, Lalana Kagal, Natasha Noy, and Eva Blomqvist, editors, *The Semantic Web – ISWC 2011*, pages 370–388, Berlin, Heidelberg, 2011. Springer Berlin Heidelberg.

[LPL⁺09] Hong Lu, Wei Pan, Nicholas D. Lane, Tanzeem Choudhury, and Andrew T. Campbell. Soundsense: Scalable sound sensing for people-centric applications on mobile phones. In *Proceedings of the 7th International Conference on Mobile Systems, Applications, and Services*, MobiSys '09, pages 165–178, New York, NY, USA, 2009. ACM.

[LPPRH10] Danh Le-Phuoc, Josiane Xavier Parreira, Vinny Reynolds, and Manfred Hauswirth. Rdf on the go: An rdf storage and query processor for mobile devices. In *Proceedings of the 2010 International Conference on Posters & Demonstrations Track – Volume 658*, ISWC-PD'10, pages 149–152, Aachen, Germany, Germany, 2010. CEUR-WS.org.

[LPSZ10] Nuno Lopes, Axel Polleres, Umberto Straccia, and Antoine Zimmermann. Anql: Sparqling up annotated rdfs. In Peter F. Patel-Schneider, Yue Pan, Pascal Hitzler, Peter Mika, Lei Zhang, Jeff Z. Pan, Ian Horrocks, and Birte Glimm, editors, *The Semantic Web – ISWC 2010*, pages 518–533, Berlin, Heidelberg, 2010. Springer Berlin Heidelberg.

[LYL⁺10] Hong Lu, Jun Yang, Zhigang Liu, Nicholas D. Lane, Tanzeem Choudhury, and Andrew T. Campbell. The jigsaw continuous sensing engine for mobile phone applications. In *Proceedings of the 8th ACM Conference on Embedded Networked Sensor Systems*, SenSys '10, pages 71–84, New York, NY, USA, 2010. ACM.

[LZD08] C. Li, Y. Zhang, and L. Duan. Establishing a trusted architecture on pervasive terminals for securing context processing. In *2008 Sixth Annual IEEE International Conference on Pervasive Computing and Communications (PerCom)*, pages 639–644, March 2008.

[Mar12] Martin Hepp. Vehicle sales ontology. http://www.heppnetz.de/ontologies/vso/ns, 2012. Accessed: May 2012.

[Mis08] J. Misic. Enforcing patient privacy in healthcare wsns using ecc implemented on 802.15.4 beacon enabled clusters. In *2008 Sixth Annual IEEE International Conference on Pervasive Computing and Communications (PerCom)*, pages 686–691, March 2008.

[MLEC07] Emiliano Miluzzo, Nicholas D. Lane, Shane B. Eisenman, and Andrew T. Campbell. Cenceme – injecting sensing presence into social networking applications. In Gerd Kortuem, Joe Finney, Rodger Lea, and Vasughi Sundramoorthy, editors, *Smart Sensing and Context*, pages 1–28, Berlin, Heidelberg, 2007. Springer Berlin Heidelberg.

[MMG11] D. McAuley, R. Mortier, and J. Goulding. The dataware manifesto. In *2011 Third International Conference on Communication Systems and Networks (COMSNETS 2011)*, pages 1–6, Jan 2011.

[MNP$^+$10] Cludio Maia, Luis Miguel Nogueira, Luis Miguel Pinho, Cludio Maia, Luis Miguel Nogueira, Luis Miguel Pinho, and Lus Miguel Pinho. Evaluating android os for embedded real-time systems. In *Proceedings of the 6th International Workshop on Operating Systems Platforms for Embedded Real-Time Applications*, 2010.

[MRS$^+$09] Min Mun, Sasank Reddy, Katie Shilton, Nathan Yau, Jeff Burke, Deborah Estrin, Mark Hansen, Eric Howard, Ruth West, and Péter Boda. Peir, the personal environmental impact report, as a platform for participatory sensing systems research. In *Proceedings of the 7th International Conference on Mobile Systems, Applications, and Services*, MobiSys '09, pages 55–68, New York, NY, USA, 2009. ACM.

[Net14] Netty. The Netty Project, homepage. http://netty.io, 2014. Accessed: August 2014.

[Nor07] Anil Nori. Mobile and embedded databases. In *Proceedings of the 2007 ACM SIGMOD International Conference on Management of Data*, SIGMOD '07, pages 1175–1177, New York, NY, USA, 2007. ACM.

[NPA10] Rammohan Narendula, Thanasis G. Papaioannou, and Karl Aberer. Privacy-aware and highly-available osn profiles. In *Proceedings of the 2010 19th IEEE International Workshops on Enabling Technologies: Infrastructures for Collaborative Enterprises*, WETICE '10, pages 211–216, Washington, DC, USA, 2010. IEEE Computer Society.

[NUI12] NUIG. Rdf-on-the-go: Triple store implementation for android. http://rdfonthego.googlecode.com/, 2012. Accessed: May 2012.

[Ord12] Ordered List Ontology. The ordered list ontology. http://purl.org/ontology/olo/core#, 2012. Accessed: May 2012.

[OT08] Tatsuaki Okamoto and Katsuyuki Takashima. Homomorphic encryption and signatures from vector decomposition. In Steven D. Galbraith and Kenneth G. Paterson, editors, *Pairing-Based Cryptography – Pairing 2008*, pages 57–74, Berlin, Heidelberg, 2008. Springer Berlin Heidelberg.

[PEC12] PECES. PECES FP7 Project, Project Homepage. http://www.ict-peces.eu, 2012. Accessed: April 2012.

[PHS10] H. Patni, C. Henson, and A. Sheth. Linked sensor data. In *2010 International Symposium on Collaborative Technologies and Systems*, pages 362–370, May 2010.

[PLA12] PLANET. PLANET FP7 Project, Project Homepage. http://www.planet-ict.eu, 2012. Accessed: April 2012.

[PPS$^+$08] J. M. Paluska, H. Pham, U. Saif, G. Chau, C. Terman, and S. Ward. Structured decomposition of adaptive applications. In *2008 Sixth Annual IEEE International Conference on Pervasive Computing and Communications (PerCom)*, pages 1–10, March 2008.

[PRAB08] L. Pareschi, D. Riboni, A. Agostini, and C. Bettini. Composition and generalization of context data for privacy preservation. In *2008 Sixth Annual IEEE International Conference on Pervasive Computing and Communications (PerCom)*, pages 429–433, March 2008.

[RB04] Philip Robinson and Michael Beigl. Trust context spaces: An infrastructure for pervasive security in context-aware environments. In Dieter Hutter, Günter Müller, Werner Stephan, and Markus Ullmann, editors, *Security in Pervasive Computing*, pages 157–172, Berlin, Heidelberg, 2004. Springer Berlin Heidelberg.

[RH10] A. Rice and S. Hay. Decomposing power measurements for mobile devices. In *2010 IEEE International Conference on Pervasive Computing and Communications (PerCom)*, pages 70–78, March 2010.

[RJH02] G. C. Roman, C. Julien, and Qingfeng Huang. Network abstractions for context-aware mobile computing. In *Proceedings of the 24th International Conference on Software Engineering. ICSE 2002*, pages 363–373, May 2002.

[RMJ$^+$11] N. Roy, A. Misra, C. Julien, S. K. Das, and J. Biswas. An energy-efficient quality adaptive framework for multi-modal sensor context recognition. In *2011 IEEE International Conference on Pervasive Computing and Communications (PerCom)*, pages 63–73, March 2011.

[RMLM09] Alejandro Rodríguez, Robert McGrath, Yong Liu, and James Myers. Semantic management of streaming data. In *Proceedings of the 2Nd International Conference on Semantic Sensor Networks – Volume 522*, SSN'09, pages 80–95, Aachen, Germany, Germany, 2009. CEUR-WS.org.

[RMM+10] Kiran K. Rachuri, Mirco Musolesi, Cecilia Mascolo, Peter J. Rentfrow, Chris Longworth, and Andrius Aucinas. Emotionsense: A mobile phones based adaptive platform for experimental social psychology research. In *Proceedings of the 12th ACM International Conference on Ubiquitous Computing*, UbiComp '10, pages 281–290, New York, NY, USA, 2010. ACM.

[RR98] Michael K. Reiter and Aviel D. Rubin. Crowds: Anonymity for web transactions. *ACM Trans. Inf. Syst. Secur.*, 1(1):66–92, November 1998.

[RSB+09] S. Reddy, V. Samanta, J. Burke, D. Estrin, M. Hansen, and M. Srivastava. Mobisense 2014 – mobile network services for coordinated participatory sensing. In *2009 International Symposium on Autonomous Decentralized Systems*, pages 1–6, March 2009.

[SAW94] B. Schilit, N. Adams, and R. Want. Context-aware computing applications. In *Proceedings of the 1994 First Workshop on Mobile Computing Systems and Applications*, WMCSA '94, pages 85–90, Washington, DC, USA, 1994. IEEE Computer Society.

[SC09] Juan F. Sequeda and Oscar Corcho. Linked stream data: A position paper. In *Proceedings of the 2Nd International Conference on Semantic Sensor Networks – Volume 522*, SSN'09, pages 148–157, Aachen, Germany, Germany, 2009. CEUR-WS.org.

[SDA99] Daniel Salber, Anind K. Dey, and Gregory D. Abowd. The context toolkit: Aiding the development of context-enabled applications. In *Proceedings of the SIGCHI Conference on Human Factors in Computing Systems*, CHI '99, pages 434–441, New York, NY, USA, 1999. ACM.

[Sem12] Semantic Desktop. Personal information model. http://www.semanticdesktop.org/ontologies/2007/11/01/pimo/#, 2012. Accessed: May 2012.

[SHL+05] Krishna Sampigethaya, Leping Huang, Mingyan Li, Radha Poovendran, Kanta Matsuura, and Kaoru Sezaki. Caravan : Providing location privacy for vanet. In *Embedded Security in Cars (ESCAR)*, 2005.

[SHS08] Amit Sheth, Cory Henson, and Satya S. Sahoo. Semantic sensor web. *IEEE Internet Computing*, 12(4):78–83, July 2008.

[SPI12] SPITFIRE Consortium. The spitfire ontology. http:// spitfire-project.eu/ontology/ns/, 2012. Accessed: May 2012.

[SPTH11] R. Shokri, P. Papadimitratos, G. Theodorakopoulos, and J. P. Hubaux. Collaborative location privacy. In *2011 IEEE Eighth International Conference on Mobile Ad-Hoc and Sensor Systems*, pages 500–509, Oct 2011.

[STD$^+$10] Reza Shokri, Carmela Troncosof, Claudia Diaz, Julien Freudiger, and Jean-Pierre Hubaux. Unraveling an old cloak: K-anonymity for location privacy. In *Proceedings of the 9th Annual ACM Workshop on Privacy in the Electronic Society*, WPES '10, pages 115–118, New York, NY, USA, 2010. ACM.

[Swe02] Latanya Sweeney. K-anonymity: A model for protecting privacy. *Int. J. Uncertain. Fuzziness Knowl.-Based Syst.*, 10(5):557–570, October 2002.

[TRL$^+$09] Arvind Thiagarajan, Lenin Ravindranath, Katrina LaCurts, Samuel Madden, Hari Balakrishnan, Sivan Toledo, and Jakob Eriksson. Vtrack: Accurate, energy-aware road traffic delay estimation using mobile phones. In *Proceedings of the 7th ACM Conference on Embedded Networked Sensor Systems*, SenSys '09, pages 85–98, New York, NY, USA, 2009. ACM.

[W3C04] W3C. N-triples specification. http://www.w3.org/TR/ rdf-testcases/#ntriples, 2004. Accessed: May 2012.

[W3C12a] W3C. Resource description framework (rdf): Concepts and abstract syntax. http://www.w3.org/TR/rdf-concepts/, 2012. Accessed: May 2012.

[W3C12b] W3C. Sparql query language for rdf. http://www.w3.org/TR/ rdf-sparql-query/, 2012. Accessed: May 2012.

[W3C12c] W3C. Sparql query results xml format. http://www.w3.org/ TR/rdf-sparql-XMLres/, 2012. Accessed: May 2012.

[W3C12d] W3C. Terse rdf triple language. http://www.w3.org/TR/ 2012/WD-turtle-20120710/, 2012. Accessed: May 2012.

[W3C12e] W3C. Time ontology in owl. http://www.w3.org/TR/ owl-time/, 2012. Accessed: May 2012.

[W3C12f] W3C. Wgs84 geo positioning: an rdf vocabulary. http://www. w3.org/2003/01/geo/wgs84_pos, 2012. Accessed: May 2012.

[W3C13a] W3C. Query results json format. https://www.w3.org/TR/ sparql11-results-json/, 2013. Accessed: May 2018.

[W3C13b] W3C. Sparql 1.1 federated query w3c recommendation. http:// www.w3.org/TR/2013/REC-sparql11-federated-query-20130321/, 2013. Accessed: June 2014.

[WCMS10] R. Wishart, D. Corapi, S. Marinovic, and M. Sloman. Collaborative privacy policy authoring in a social networking context. In *2010 IEEE International Symposium on Policies for Distributed Systems and Networks*, pages 1–8, July 2010.

[Wei91] Mark Weiser. The computer for the 21st century. *Scientific American*, 265:94, 09 1991.

[WLA$^+$09] Yi Wang, Jialiu Lin, Murali Annavaram, Quinn A. Jacobson, Jason Hong, Bhaskar Krishnamachari, and Norman Sadeh. A framework of energy efficient mobile sensing for automatic user state recognition. In *Proceedings of the 7th International Conference on Mobile Systems, Applications, and Services*, MobiSys '09, pages 179–192, New York, NY, USA, 2009. ACM.

[WZL06] Kamin Whitehouse, Feng Zhao, and Jie Liu. Semantic streams: A framework for composable semantic interpretation of sensor data. In Kay Römer, Holger Karl, and Friedemann Mattern, editors, *Wireless Sensor Networks*, pages 5–20, Berlin, Heidelberg, 2006. Springer Berlin Heidelberg.

[XC09] Toby Xu and Ying Cai. Location safety protection in ad hoc networks. *Ad Hoc Networks*, 7(8):1551 – 1562, 2009. Privacy and Security in Wireless Sensor and Ad Hoc Networks.

[YTN05] T. Yamabe, A. Takagi, and T. Nakajima. Citron: a context information acquisition framework for personal devices. In *11th IEEE International Conference on Embedded and Real-Time Computing Systems and Applications (RTCSA'05)*, pages 489–495, Aug 2005.

[ZH09] G. Zhong and U. Hengartner. A distributed k-anonymity protocol for location privacy. In *2009 IEEE International Conference on Pervasive Computing and Communications*, pages 1–10, March 2009.

[ZS12] Stefan Zander and Bernhard Schandl. Context-driven rdf data replication on mobile devices. *Semant. web*, 3(2):131–155, April 2012.

Index

About the Authors

Dr. Marcus Handte received a master's degree in computer science from the Georgia Institute of Technology in 2002 and his Diploma in Software Engineering from the University of Stuttgart in 2003. From August 2007 to October 2009 he was working as a full-time researcher at Fraunhofer IAIS, since then he is working at the University of Duisburg-Essen. In 2009 he received a PhD in computer science from the University of Stuttgart and in 2013 he received his postdoctoral lecture qualification from the University of Duisburg-Essen. His past research focused on middleware for adaptive and self-configuring systems. From 2012 to 2015, he was responsible for the technical coordination of the research and development work within the GAMBAS project.

Prof. Dr. Pedro José Marrón received his bachelor and master's degree in computer engineering from the University of Michigan in Ann Arbor in 1996 and 1998 and his Ph.D. from the University of Freiburg in 2001. After a professorship at the University of Bonn, he is currently full professor at the University of Duisburg-Essen, where he leads the "Networked Embedded Systems Group". Pedro Marron is also founder of Locoslab GmbH, an SME specialized in low cost solutions for localization in indoor environments and is also the president of UBICITEC, the European Center for Ubiquitous Technologies and Smart Cities, which counts with over 20 institutional partners from industry and academia.

Prof. Dr. Gregor Schiele is a full professor for computer science and the leader of the research group for embedded systems at the University of Duisburg-Essen, Germany. His research focusses on adaptive systems and the Internet of Things (IoT). Before joining the University of Duisburg-Essen in 2014, Dr. Schiele worked in different roles at the National University of Ireland in Galway, the Digital Enterprise Research Institute (DERI) and the

Insight Centre for Data Analytics in Ireland, as well as at the Universities of Mannheim and Stuttgart in Germany. He received his doctorate in computer science from the University of Stuttgart in 2007 for his work on System Support for Pervasive Computing. Dr. Schiele has published over 50 papers, articles and book chapters. He served in more than 100 international technical program committees, is a reviewer for multiple international journals and co-organised a multitude of scientific workshops and conferences. His research received funding from the DFG, the DAAD, a multitude of industry partners, as well as the EU. He is collaborating closely with partners in the EU and worldwide, including in Norway, South Africa, and the USA and is a regular reviewer for international journals and research programs.

Manolo Serrano Matoses is an Electronic Engineer. He holds a double degree in Telecommunications from the Polytechnic University of Valencia (Spain) and l'École Nationale Supérieure des Télécommunications de Bretagne (France) – networking specialisation. He is currently Head of the New Technologies Area within the Technology department of ETRA I+D. Manolo actively participated in the GAMBAS project, and coordinated the EMMA and PECES projects, predecessors of the work done in GAMBAS. Manolo is passionate about technology, and he is responsible within its company on the proposal of new innovative paths to improve the company products and explore new business application domains.